GIS
in
Telecommunications

Lisa Godin

ESRI PRESS

REDLANDS, CALIFORNIA

ESRI
GIS in Telecommunications
ISBN 1-879102-86-2

First printing May 2001. Second printing November 2001.

Printed in the United States of America.

Library of Congress Cataloging-in-Publication Data
Godin, Lisa Winget, 1972–
    GIS in telecommunications / Lisa Winget Godin.
        p.   cm.
    ISBN 1-879102-86-2
    1. Artificial satellites in telecommunication.
2. Telecommunication. 3. Geographic information systems. I. Title.
TK5104.G62 2001
621.382—dc21   2001002629

Published by ESRI, 380 New York Street, Redlands, California 92373-8100.

Books from ESRI Press are available to resellers worldwide through Independent Publishers Group (IPG). For information on volume discounts, or to place an order, call IPG at 1-800-888-4741 in the United States, or at 312-337-0747 outside the United States.

Contents

Other books from ESRI Press

Acknowledgments

A large number of people were involved in making this book happen. Those from the organizations involved who shared their applications and reviewed the material for accuracy are listed at the end of each case study. This book could not have been written without them.

Many people from ESRI also contributed their time and expertise. Kees Van Loo helped identify and keep track of the case-study candidates, a tough job in an industry so volatile. Michael Karman edited the book while also acting as a sounding board whenever I needed it. Gary Amdahl and R. W. Greene also took time out of their busy schedules to edit the book, offering a fresh perspective through their editorial comments.

Jennifer Galloway did the page layout and production. Michael Hyatt designed the book and did the copyediting. Doug Huibregtse designed and produced the cover. Cliff Crabbe oversaw print production. Randy Worch offered counsel when I had obscure technical questions. Barbara Shaeffer reviewed the final manuscript.

Christian Harder provided both technical and human resources and Judy Boyd and Nick Frunzi set high quality standards.

Chris, Michael, Chance, and Riley gave up countless hours with me, quietly (and sometimes not so quietly) offering moral support and inspiration.

Finally, special thanks to ESRI President Jack Dangermond, for recognizing the value case-study books offer in educating the world about the power of GIS.

Lisa Godin
Redlands, California

GIS for the information age

The Telecommunications Act of 1996 set the stage for healthy competition between telecommunication companies. The increasing popularity of mobile communication devices, aggressive growth strategies, and the continuing rise of the Internet have since turned that healthy competition into a global rivalry.

Aggressive growth means competing with companies all over the world. As more and more telecommunication companies move from serving a city, state, or region to serving the entire globe, a universally insatiable appetite for information has prompted them to find ways to provide better, faster, and less expensive service to their customers.

The use of GIS for telecommunications started with applications for automated mapping and facilities management. In the wireless industry, the first applications were focused on providing support for wave propagation analysis.

While engineering applications remain crucial to any mature implementation, GIS has expanded from a single-department application to a companywide endeavor. Attention has shifted to sharing data between departments, building the infrastructure that makes delivering information via the Internet and intranets possible. With

GIS technology following IT standards and increasing performance capabilities, more and more telecommunications providers are establishing and honing a competitive edge by integrating their workflow based on the location of their assets, customers, sales territories, and coverage areas.

As telecommunication networks grow more complex, the GIS applications they use must become more sophisticated.

Expanding networks

When it comes to telecommunication companies, GIS has traditionally been used to manage spatial information as it relates to outside plant facilities—those parts of a company's network that exist out in the field. No longer limited to the traditional telephone poles and lines, today's telecommunication networks include everything from fiber-optic cables to radio antennas to cellular base stations and all the parts in between.

Regardless of the number or types of features involved in a particular network, ESRI® GIS software lets engineers add or delete features as necessary, trying different designs until they find the most suitable one. All this with point-and-click technology that lets them quickly perform analyses that might otherwise take days to complete.

Deciding where to expand its network and which services to provide are two of the most important decisions a telecommunication company must make. The success of these decisions depends on providing the right services to the right people at the right time.

Locating potential customers and deciding which services to provide them involves studying data that often comes from other departments or even from outside the company.

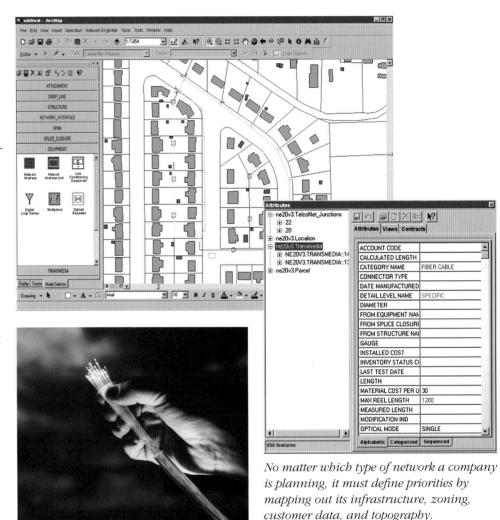

No matter which type of network a company is planning, it must define priorities by mapping out its infrastructure, zoning, customer data, and topography.

Integrating data

More than one department is involved in expanding a network, from those that provide data to those that approve work orders to those that actually go out and physically install the facilities. To be efficient, changes to the network must be made nearly as fast as their need is identified, a process that is impossible if each department has to access, process, and review data stored in separate databases. This is complicated further when the various departments use different software to manage the data. GIS solves this problem by making it easy to process, store, and retrieve data in a centralized database.

Storing data in a centralized database ensures that users throughout the company have constant access to up-to-date information, whether that information came from departments within the company, like customer service, marketing, or engineering, or from outside the company, such as census, land-use, or topographical data.

ESRI's ArcSDE™ makes it easy to integrate and manage spatial data in a database management system like IBM® DB2®, Informix®, Microsoft® SQL Server™, or Oracle®. ArcSDE serves spatial data to the ArcGIS Desktop suite (ArcView®, ArcEditor™, and ArcInfo™) and through ArcIMS®, as well as other applications, making it the key component in managing a multiuser spatial database.

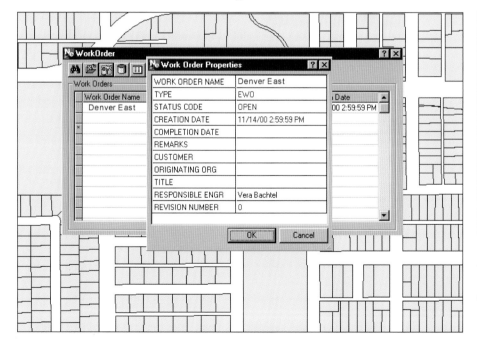

The fastest and easiest way to gain market share is to offer a service before competitors do. Enterprise GIS enables you to do just that.

Serving customers

No matter how carefully a company builds its network, it won't last long in the telecommunication industry without good customer service.

Customer service departments frequently take calls from customers who want to know if service is available in their area or who may have had disruptions in their service from a network outage. With access to a centralized database, customer service representatives are better able to provide the customer with accurate information about service availability or how long an interruption in service is likely to last.

Provisioning is the single most important piece in telecommunications GIS. A better provisioning process with state-of-the-art tools, taking advantage of accurate geographic information, is the key to maintaining a competitive advantage.

Providing customers with the information they need when they need it is crucial for success. The more accurate the information, the happier the customer.

Sharing the info

The popularity of the Internet has encouraged most companies to share information on the Web. Whether they elect to share information with employees through their own intranets, with customers and clients through extranets, or with the general public through the Internet, GIS helps make it happen. ESRI's ArcIMS is the only software that enables users to integrate local data sources with Internet data sources for display, query, and analysis in an easy-to-use Web browser. ArcIMS puts a world of information on your desktop by simultaneously accessing Web data, local shapefiles, SDE® layers, and images for viewing with local data.

ArcIMS contains wizards and templates that help guide you through tasks for authoring and publishing maps—no programming is required. Simply create a map service, design the Web pages, and publish. Easy-to-use tools help monitor and maintain the site. Client and server configuration and management tools are also available for building secure, reliable, and highly scalable sites.

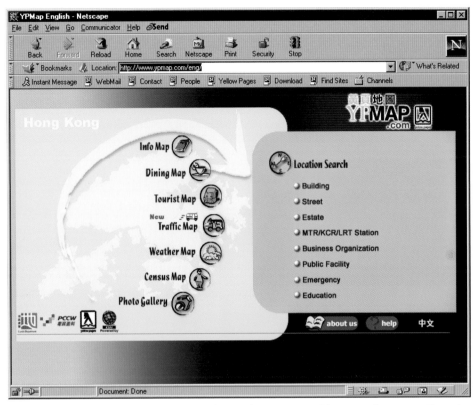

Using ArcIMS, ArcSDE, and ArcInfo software from ESRI, Cable & Wireless Hong Kong is able to offer a wealth of information on the Internet.

Location is everything

The instant information of the Internet combined with the escalating popularity of Web-enabled mobile communication devices like cellular phones, two-way pagers, and handheld PCs have raised the stakes for telecommunication companies. Customers demand instant information and consistent service regardless of where they are. The struggle to compete for information-hungry and increasingly mobile customers has spawned a whole new branch of services, those based on location.

Location-based services enable carriers and their business partners to offer unprecedented services to mobile subscribers. More and more mobile devices are being designed to exchange information with GPS or other positioning technologies. Users can look for the nearest gas stations, restaurants, or pharmacies, and get driving directions to the location.

Location Sensitive Billing allows wireless carriers to offer multiple rate zones to their customers. Wireless customers will typically pay less in their home zone, with differing rates for other zones of service. With GIS, wireless carriers can instantly identify which zone a call originates in. Location Sensitive Billing is just one GIS-based service that allows wireless carriers to compete with local and long-distance landline carriers.

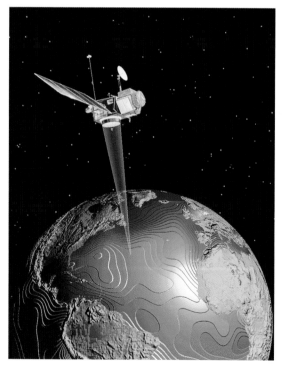

With the convergence of wireless communication, location determination technologies such as GPS, and personal digital assistants (PDAs), the need for the tools that GIS technology provides grows stronger every day.

Finding what you want

Getting from A to B without knowing the local roads and freeways can be a challenge. Location services allow travelers to find their way, no matter where they are or where they want to go. And if they get lost, all they have to do is enter their destination, and GIS will help them get back on track.

Traffic conditions change continually; traffic information, in real time, is useful when and where people need it. Subscribers can now get live traffic information, along with maps and alternate routes, using cellular phones and other mobile devices. When subscribers find out about congestion along their routes, GIS location services can immediately provide alternate routes.

The same technology lets wireless carriers pinpoint the location of a cellular phone in case of an emergency 911 call. The cellular phone location is then entered into the GIS, and the appropriate emergency response team is routed to the caller's position.

Even if it's not an emergency, locating a mobile device can come in handy. Whether customers want to locate their cars, their kids, or their coworkers, the combination of GPS-enabled mobile devices and GIS makes it possible.

As the workforce becomes increasingly mobile, location services will not only let companies keep track of their employees, it will also enable personnel to access remote data servers with portable, wireless devices. This way workers in the field will have access to the information they need, and enable real-time updates to corporate databases.

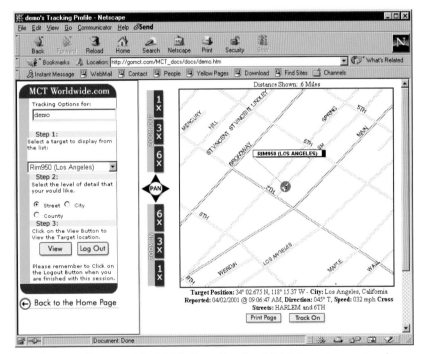

Tracking applications like MCT's PortaTrack, which lets customers track just about anything, would not be possible without the convergence of GPS technology and GIS.

The case studies

This book demonstrates how various telecommunication companies use GIS to manage their networks, provide better customer service, and create new and exciting services.

2

It's all about where you are

Mobile phones were designed to let people communicate no matter where they are. However, where people are does matter. It matters to their employers, it matters to their families, and it matters to the people using the phones. This apparent contradiction is the key to the newest types of service to hit the wireless industry—those based on location.

This chapter will introduce you to SignalSoft Corp.'s Wireless Location Services®, a GIS-based product suite that allows wireless network operators to take full advantage of one of the most important aspects of mobile phone service—the location of their subscribers.

Wireless Location Services not only lets network operators determine the latitude and longitude of a wireless call, it provides valuable information about the surrounding area; whenever the subscriber makes a call or enters a code, the system compares that location with a database of information about the surrounding area. This can translate into myriad benefits: the price of calls made from within a specific area can be cheaper; more precise vehicle tracking can ensure better delivery service; and the instantaneous location of a 911 call can even mean the difference between life and death.

SignalSoft's Wireless Location Services applications include Location Sensitive Billing, Wireless 911, local.info, and BFound™, an Internet tracking application.

Sending signals

When a subscriber initiates a wireless call, the handset sends out a radio frequency (RF) signal. The signal arrives at cell site after cell site until it becomes too weak to travel any farther. Each cell site captures information about the signal, like the wireless phone's electronic serial number, and the signal's strength, time of arrival, and direction.

Although this data is stored at the cell site, it doesn't go to the Wireless Location Services applications automatically. In most cases today, the user has to press a button or enter a code to invoke the Wireless Location Services system.

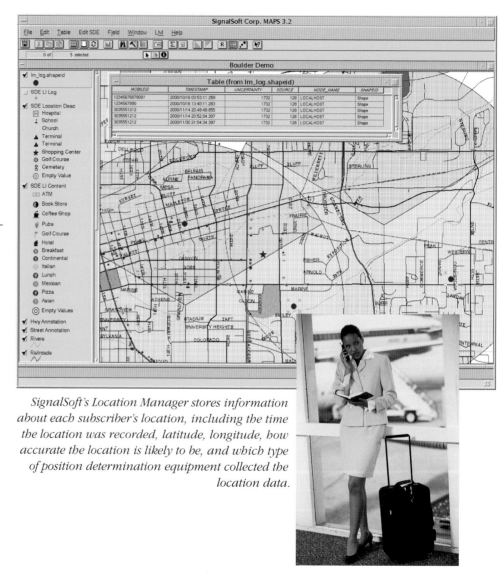

SignalSoft's Location Manager stores information about each subscriber's location, including the time the location was recorded, latitude, longitude, how accurate the location is likely to be, and which type of position determination equipment collected the location data.

Seeing it on the map

When a subscriber makes a call or enters a code, a Wireless Location Services application called Location Manager records the phone's electronic serial number and then retrieves the location, time of call, and signal direction. Location Manager converts the location into a pair of latitude and longitude coordinates, which it then returns to the application, along with a calculation of how certain it is that the location is accurate—a mathematical calculation often performed by the position determination equipment itself, in much the same way that the margin of error is calculated by a GPS device.

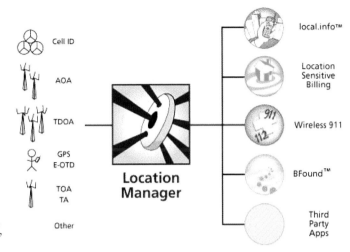

Location Manager provides interfaces that allow access to every type of position determination equipment (PDE), an advantage that eliminates the decision between PDE and Wireless Location Services applications. Without these built-in interfaces, wireless carriers would have to go to the PDE provider to create a new interface every time they wanted to add a new application.

Saving money

Subscribers can be charged less for calls made in certain areas, a benefit called Location Sensitive Billing (LSB).

Network operators set up rate zones for each subscriber. Once Location Manager returns the subscriber's location, the Location Sensitive Billing application compares that location with the subscriber's rate zones.

Subscribers don't need to program their phones or turn the service on or off. Simply subscribing to the Location Sensitive Billing service means that the subscriber's location will automatically be requested and returned by Location Manager.

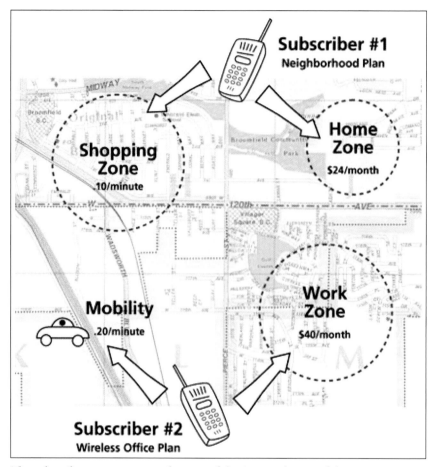

The subscriber can enter a code to see if they're inside one of their rate zones before they make a call.

Keeping it straight

Each subscriber's rate zone is created with an application known as MAPS (Mobile Application Provisioning System). Network operators use the same application to view and modify the rate zones and to store and manage the location information coming in from the position determination equipment.

MAPS lets network operators manage their spatial data in an Oracle database, making it easy to keep the data current and available.

If a customer calls to ask why she was billed the way she was for a particular call, the customer service representative uses MAPS to bring up a map of where the call originated, letting the customer know whether that location is in any of her rate zones.

While Location Sensitive Billing lets subscribers save money, the concept behind SignalSoft's Wireless Location Services could even save their lives.

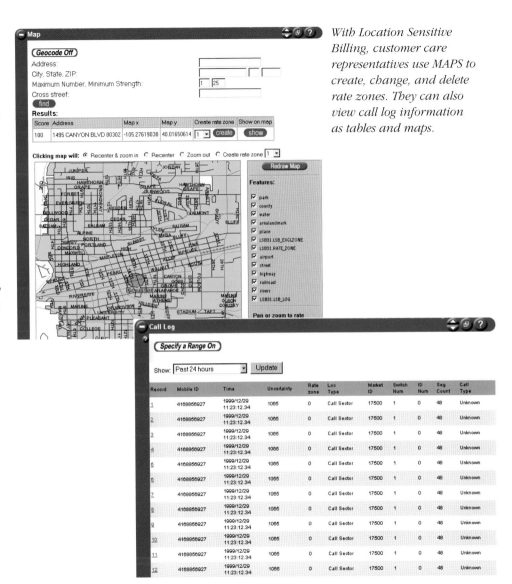

With Location Sensitive Billing, customer care representatives use MAPS to create, change, and delete rate zones. They can also view call log information as tables and maps.

Getting help when you need it

You didn't even see the accident coming. The next thing you know you're stuck in a ditch 100 feet below the road. Your car is wrecked and you're pinned behind the wheel. At least you can still feel your legs. Fortunately, your service provider offers Wireless Location Services, which means 911 operators will find you even if you don't know where you are.

Wireless subscribers make more than 25 percent of all 911 emergency calls—more than eighty thousand calls every day. With the increasing popularity of wireless phones, this number will only increase. Until recently, however, there was no way for emergency dispatchers to know where calls from a wireless phone came from.

In June 1996, the Federal Communications Commission ordered wireless service providers to implement Enhanced 911 (E911) technology. Wireless carriers first had to provide emergency dispatchers with a wireless caller's number and the location of the nearest cell site. And, beginning in October 2001, carriers will have to identify the location within 125 meters at least 67 percent of the time. Signal-Soft's Wireless 911 application lets network operators comply with both parts of the FCC order.

With SignalSoft's Wireless 911 application, network operators can not only find the latitude and longitude of a person calling 911, but transform that location into an address, and also route the call to the correct emergency response service.

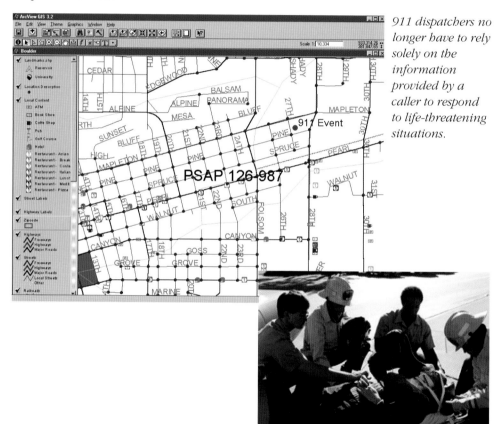

911 dispatchers no longer have to rely solely on the information provided by a caller to respond to life-threatening situations.

Keeping track of the important things

SignalSoft's Wireless Location Services can not only pinpoint locations, but with the acquisition of BFound.com, a leading developer of Internet-based tracking technology, SignalSoft now offers sophisticated tracking software.

Using only a standard Web browser and a mobile phone, subscribers can track packages, see when their children leave school and arrive home, or manage vehicles and sales teams as they move through the field. Various security measures, such as password protection, ensure that BFound accounts cannot be monitored by third parties.

Tracking people used to be more limited—a GPS unit in a car's trunk could track the car, but not the person. Because today's technology is carried by a cooperating person, it allows more precise personal tracking to be available to anyone with a wireless phone.

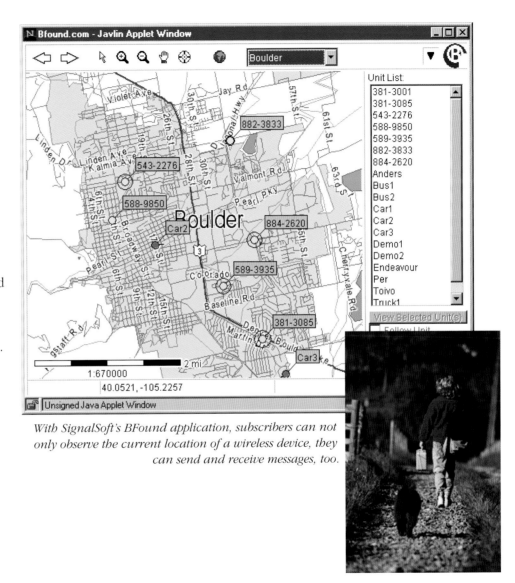

With SignalSoft's BFound application, subscribers can not only observe the current location of a wireless device, they can send and receive messages, too.

Finding out what's happening around you

Subscribers can also use wireless phones to get up-to-the-minute information about where they are. Using local.info™, another part of SignalSoft's Wireless Location Services, subscribers can find out what traffic will be like on the drive home, learn where the nearest hotel or restaurant or gas station is, or be instantly connected with roadside assistance.

As with the other Wireless Location Services applications, when a subscriber requests information from local.info, the system locates that subscriber, then translates the location to a zone or business description. The local.info application then retrieves the appropriate information from the companies that provide information for the system. This information may reside within the local.info database or on the Internet or some other server outside the application. SignalSoft's Remote Content Interface retrieves this data with XML and then sends the data directly to the user.

SignalSoft's customers can also use MAPS to set the number of services or businesses the system returns to a subscriber using local.info. Limiting the number allows network operators to be more specific, giving subscribers only the information that most closely matches their request. A subscriber looking for the nearest ATM will want to know where the five closest ATMs are, not the fifty closest.

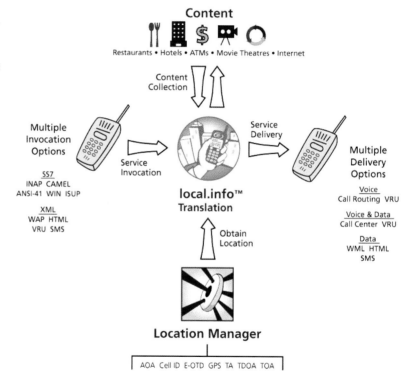

Local.info also lets you access the Internet and even shop online. The information can be accessed via Wireless Application Protocol (WAP), voice recordings, short messages, or live through call centers.

Looking forward

In the near future, some Wireless Location Services applications may not have to be specifically invoked. For example, users may opt to have local.info automatically inform them of traffic conditions for a specific area once they get within a certain distance of it. An art lover who periodically visits a large city some distance from home may want to know about featured art exhibits when in town. Parents may automatically be notified the moment their child leaves school grounds or arrives home. This combination of location technology and GIS is the heart of SignalSoft's Wireless Location Services, allowing SignalSoft to offer network operators more ways to serve their customers than ever before.

With the wide variety of applications, many of which have yet to be fully realized, location services are likely to affect our lives even more than the wired Internet has.

The system

Hardware: SignalSoft's applications can operate on either service control points or on non-SS7/IN platforms such as Sun™.

Software: ArcView, SDE, ArcIMS

The data

All represented basemap data sourced from public and/or commercial third-party geodata suppliers.

Location-enabling cell and RF data granted by Network Operators.

Location-sensitive content sourced from SignalSoft Content Alliance partners.

Other application-specific ancillary data sourced from SignalSoft.

The people

Thanks to Angie Emery, communications manager; Jonathan Spinney, MAPS product specialist; and Gerry Christensen, director of product management, all of SignalSoft Corp.

Keeping track of things

3

Tracking where people and things are on the globe, and where they are going, is becoming an easier and more commonplace task. Devices such as GPS receivers and cellular telephones, to mention only a couple of the tools of tracking technology, get smaller, cheaper, and easier to use every day, expanding in turn the application capabilities of tracking technology.

Managing the data returned from such devices requires tracking software such as PortaTrack by Main Course Technologies, Inc. (MCT). In this chapter you will see how this company uses GIS to help you keep track of the things you care about most.

Founded in 1998, MCT, of Los Angeles, California, prides itself on the technology it has developed and on the solutions its technology provides.

MCT's PortaTrack application is a tracking service that uses ESRI's ArcIMS to seamlessly integrate wireless devices

and Internet applications. Through a tracking and wireless data transmission device, such as a cell phone, PortaTrack customers can view on a map the location of a vehicle, person, or any other mobile object on a map.

Specifically built to serve GIS on the Internet, ArcIMS is designed to make it easy for companies such as MCT to create map services, develop Web pages for communicating with the map services, and administer sites like PortaTrack.

How it works

When a GPS device receives signal information—such as position, time, date, direction of travel, and speed—from a wireless device such as a cell phone, it sends that information over the wireless network. PortaTrack receives this information, combines it into a map, then provides it to customers over the Internet—accessible, of course, through a desktop or laptop PC, Web-enabled phone, or interactive pager.

The PortaTrack application is composed of several tightly integrated applications in a distributed architecture. Data servers collect satellite, cellular, or other wireless location data and serve the information back to the tracking databases. These databases store this information, along with specific characteristics of each particular device—for example, its transmission capability, that is, its ability to report its position, either remotely or based on intervals of time or mileage.

The PortaTrack application retrieves and displays location data from the tracking databases. Based on ArcIMS, PortaTrack can build street-level detail maps with as many as sixteen layers of mapping and tracking data.

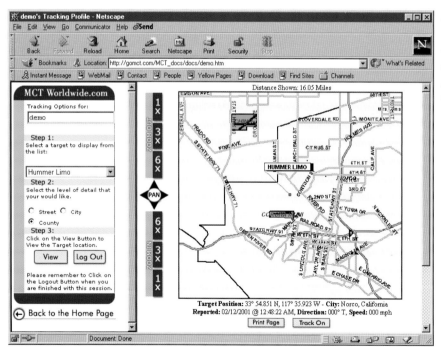

Below each map, PortaTrack displays information about the date and time of the last position report; latitude and longitude; street and cross-street position; city and state; and speed and direction at the time of the last reported position.

Keeping it simple

While all this sounds complicated, PortaTrack is easy to use. Just sign on, pick a target, and read the map. There are no software applications to buy or download to the client computer, so there is no waiting to see the data. It is a stable and reliable service on the widest variety of client platforms and browsers.

MCT designed PortaTrack to receive tracking and data information from a wide variety of sources because no data transmission system is perfect for every use. Even different types of wireless devices can be used on the same account. Each account is username- and password-protected so the data is only available to those with proper authorization. Any authorized person can see the data on any Web browser with an Internet connection.

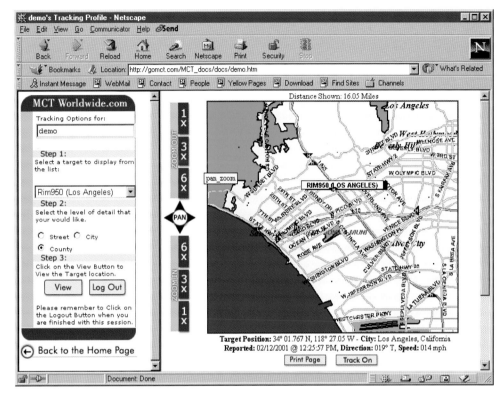

After the hardware is installed and services are requested, access to the information is usually available within minutes of installation.

Minding your business

The technology can have wide applicability for many kinds of businesses, large and small. For example, suppose you run a small locksmith business that needs some help to streamline efficiency and allocation of resources. At present, when a service call comes in, you may send the customer's phone number and address via pager to the service technician responsible for that area and let the service tech determine his own ETA and priority for the call. Or you may send out a broadcast alert to all technicians in the area, and hope they coordinate the ETA and priority among themselves. Neither solution is working very well. There is no way to see where the service technician actually is or to tell if another technician is closer to the customer. In fact, the office only gets updates from the service technicians when they call in for parts or when they reschedule a job. With technicians dispatched according to the areas they cover rather than their proximity to the call, customers sometimes have to wait longer than necessary for help to arrive. The result is you're losing business.

Once you sign up for PortaTrack and have the systems installed, all you have to do is log in to the PortaTrack Web site and choose which truck you want to see. A map appears showing that truck's current location.

When a customer calls, you can enter the customer's address into the PortaTrack system and immediately see which truck is closest to the destination. With the customer still on the line, you can page the customer's location to the nearest driver, who pages back an ETA so you can let the customer know that help is on the way—all in a matter of moments.

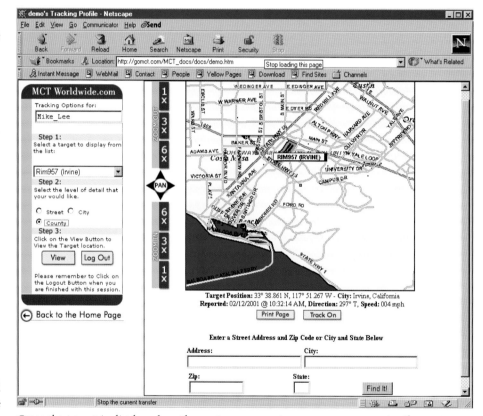

Once the target is displayed on the map, you may pan, zoom, or recenter the map. Entering a location lets you see whether a particular object is near its destination.

Playing it safe

Booms in the construction industry mean that demand for heavy equipment often exceeds the supply. Since one key fits thousands of bulldozers, backhoes, or dump trucks, construction equipment is an easy target for thieves who see the shortage as an opportunity to make some easy money.

With only 10 percent of stolen construction equipment ever recovered, rising insurance premiums pressure owners to better secure their equipment. Temporary fencing around sites is no match for a ten-ton truck, and paying a guard to stand watch at each site is prohibitively expensive.

PortaTrack offers construction foremen an affordable way to make sure that the equipment they need will be there when they need it. Not only can they keep an eye on expensive equipment at risk of theft, they can also see where equipment is if it doesn't arrive at a job on time.

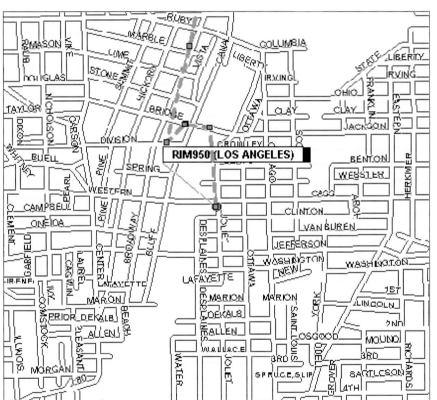

The Track On/Off button overlays up to thirty previously recorded positions, or track points, on the map. By choosing "Track On" from the PortaTrack interface, a foreman waiting for a backhoe can follow its path, estimate its average speed, and determine how far it still has to go before it arrives at its destination.

Bringing it home

PortaTrack is not limited to commercial use; you can also use it at home. For example, suppose your daughter is about to get her driver's license and you want to make sure that she's safe behind the wheel. You decide she can have her own car, but only if you install a black box. This device can transmit information about a vehicle's location, as well as how fast it was going, in what gear, and for how long. Not only will it tell you where she is, it will tell you whether or not she is driving safely.

Shortly after installing the system, she doesn't come home on time. You log on to PortaTrack and find her within seconds. She's stuck on the freeway with a flat tire, and you go rescue her.

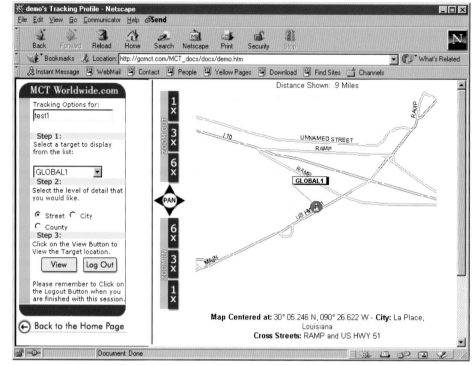

Black boxes not only let parents know where their kid is, but whether they've been playing by the rules.

Coming in out of the cold

Even if they don't drive, kids can carry wireless phones or handheld PCs or pagers so parents can keep track of them wherever they go. Logging on to the PortaTrack Web site from their computer, wireless phone, or other Web-enabled device lets parents immediately locate their child.

School districts that have PortaTrack installed on school buses can allow parents and administrators to track buses along their routes, or when they take kids to off-campus events like field trips or sporting competitions. Parents can even log on in the morning to see when the bus is on its way, ensuring that kids don't miss the bus or spend too much time waiting out in the cold.

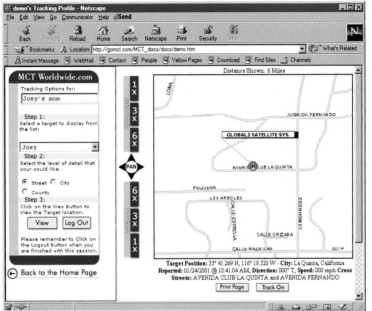

Web-enabled devices used for tracking, like wireless phones, handheld PCs, or pagers, often let people communicate with each other as well.

Exploring the possibilities

Some elder-care providers use Porta-Track to monitor high-risk patients, like those with Alzheimer's disease, who may wander away from home. These patients wear a portable GPS device, similar to a pager, on their belts.

The nursing staff can log on to the PortaTrack Web site to see where their patients are, while those who leave to retrieve lost patients can monitor the patient's location using a cell phone or handheld PC.

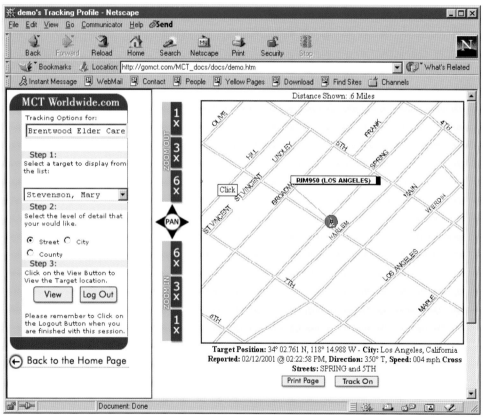

When patients who require medical care wander off, the time it takes to find them can make the difference between life or death. With PortaTrack, by the time a patient turns up missing, she's been found already.

Watching from a distance

Some great long-distance races inspire the imagination, events like the Iditarod dogsled race, the Transpacific and Whitbread around-the-world sailboat races, and the Baja 1000 off-road race, for instance. Impossible to watch, fans have long been denied even a glimpse of the action. By giving a tracking device to each participant, event officials and spectators alike can keep track of the event, every step of the way.

By embedding MCT's Web-site tracking in their Web site, sponsors offer fans a way to connect with remote events they might otherwise miss, generating enthusiasm and revenue along the way. Remote tracking of these kinds of events often rivals even the best television coverage since all too often the media is only able to speculate as to the progress of such an event.

Photo courtesy of Jeff Hensen

No matter where they are traveling, or playing, PortaTrack tracking software can help people stay in sight.

The system

Hardware: PortaTrack works with any device that has a GPS and wireless connection, including cellular phones with GPS receivers, two-way pagers, AVL transponders, and similar devices.

Software: ArcIMS

The data

Maps are created with data from the ESRI Data & Maps CD–ROM.

The people

Thanks to Patrick Coggins and Peter Woodward, both of Main Course Technologies.

Location at your fingertips

4

The Qwest for perfection

Thousands of people a day call their telecommunication companies with service questions. Sometimes it's people planning a trip who want to know if they can use their phones the whole way. Sometimes it's people on their way to work who inexplicably experience a dropped call. Sometimes the phone itself needs to be repaired. Whatever the reason, customer representatives work long and hard to keep customers connected.

In this chapter you will see how Qwest Communications' wireless division, Qwest Wireless, uses GIS to improve customer service, focus its marketing efforts, and even fight crime.

the area it covered changed frequently, sometimes faster than these traditional databases could be updated. Not only did this make it difficult to provide customers with up-to-date information, but representatives could only see the information in table and chart form.

In 1997, Qwest Wireless solved this problem by implementing a GIS. It developed a customized extension of ArcView GIS called Wildcat (Wireless Coverage Analysis Tool) for use throughout its company.

For years, hundreds of customer service representatives at Qwest Wireless answered thousands of customer inquiries a day with information from a set of related databases. The capacity of the Qwest Wireless network and

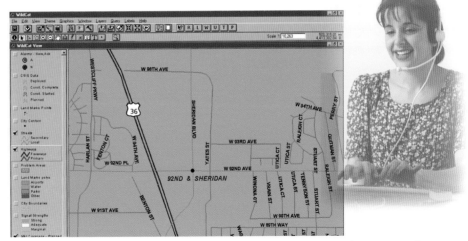

With Wildcat, not only can representatives provide customers with current information, they can also see the information geographically.

Looking for coverage

A couple in Denver, Colorado, calls Qwest's Wireless Customer Care Center to see if they will have coverage when they take a ski trip to Vail, Colorado. Using Wildcat, the customer service representative performs a query on service coverage between Denver and Vail. Wildcat then returns a map showing where the couple will have coverage. In a matter of seconds the representative can tell the couple that not only will they have coverage while in Vail, but also while traveling between the two towns.

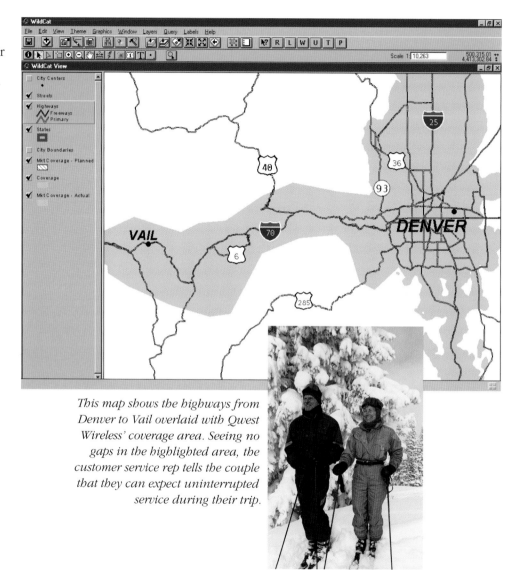

This map shows the highways from Denver to Vail overlaid with Qwest Wireless' coverage area. Seeing no gaps in the highlighted area, the customer service rep tells the couple that they can expect uninterrupted service during their trip.

Fixing the problem

So now our couple is skiing happily down the slopes, cell phones fully operational wherever they go. Meanwhile, you're stuck back in the city, on the road and on the phone to your office, when the phone goes dead: a dropped call. Once back at your desk, you call the Wireless Customer Care Center. The representative asks you where exactly your call was dropped. Entering that location into Wildcat returns a coverage map that shows known problem areas. As it turns out, your call was dropped in an area Qwest Wireless knows about and is working on fixing.

By clicking on the problem area on the screen, the representative can see that the problem is due to be fixed by the end of the month. You're mollified, somewhat: at least somebody is working on it. You turn back to your work, wishing you could spend the day somewhere else—skiing in the mountains, perhaps.

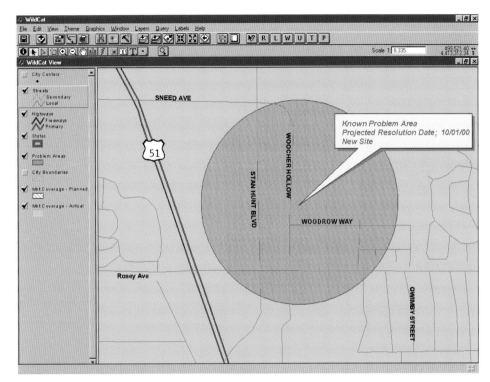

Maps like the one above make it easy for customer service reps to inform customers about how widespread a problem is and how long it will last.

When phones break

So now you're happy, but others are not. The customer service rep is already on the phone again, helping a customer having problems with a wireless phone. Unable to activate the handset, the customer would like to know where to go to have it repaired.

Customer service representatives at Qwest Wireless use a GIS application called GWIZ to see the locations of sales and service locations on a map of the customer's area. Simply entering the customer's location brings up a map of nearby service centers.

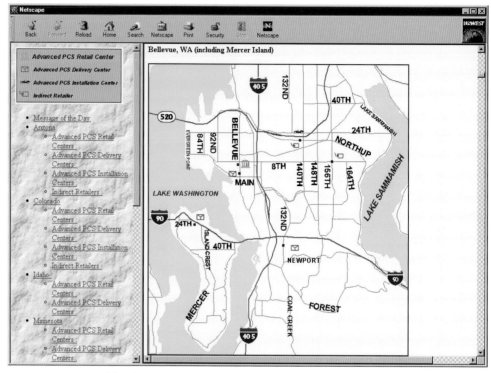

By viewing a map like this one, which shows retail, delivery, and installation centers, representatives can provide customers with information about where they can buy new equipment or have existing equipment serviced.

When all else is well

Sometimes the problem is neither service availability nor problems with a customer's phone. Sometimes the problem is at a base station—for example, a mechanical breakdown or lightning strike that makes the station temporarily unable to provide coverage.

Employees in the Network Operations Center use Wildcat to access coverage information so they can see where base stations are and which areas they cover. If a base station has problems, Wildcat automatically shows which areas are affected. Reps can then assure the customer that they know what the problem is and are fixing it.

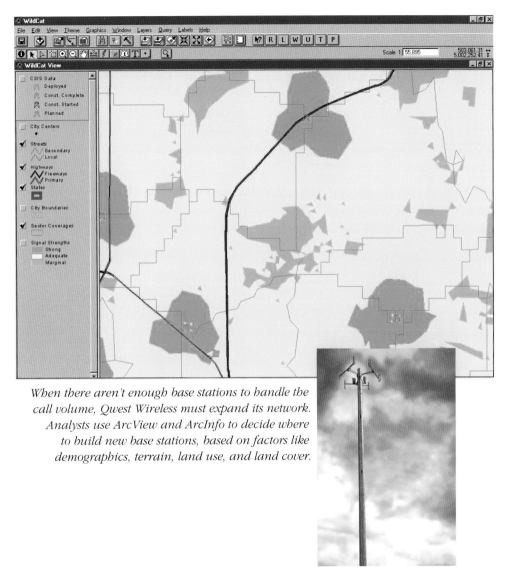

When there aren't enough base stations to handle the call volume, Qwest Wireless must expand its network. Analysts use ArcView and ArcInfo to decide where to build new base stations, based on factors like demographics, terrain, land use, and land cover.

Detecting fraud

Occasionally customers complain that they didn't make a call listed on their monthly statement. When this happens, Qwest Wireless must figure out whether a crime has been committed.

For years one of the most popular types of wireless fraud was cloning—programming the electronic serial number and the mobile identification number assigned to one phone into another phone. The second phone is able to make calls, but the bill for the airtime goes to the first one.

Although cloning occurs most often in analog networks, customers with digital phones are not safe from other types of fraud. People sometimes steal phones, use someone else's phone in an area served by another carrier (accumulating roaming charges along the way), or sign up for service under a false identity.

To combat these kinds of fraud, analysts at Qwest Wireless use GIS to compare the cell sector and cell boundary layers in Wildcat with billing records. By paying special attention to where its customers have been and what their regular calling habits are, Qwest Wireless is often able to detect fraudulent activity before the customer even suspects that something is wrong.

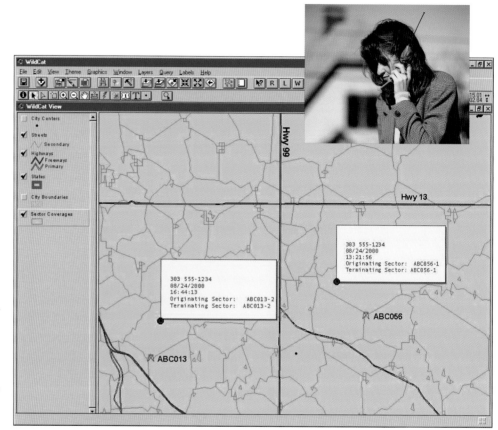

Using maps that update in real time, Qwest Wireless can sometimes spot a fraudulent call even while it is occurring. Analysts can see a call from one billing number and then only a minute later, a call from many miles away using the same number.

Narrowing it down

Qwest Wireless uses GIS not only to help current customers, but to gain new ones as well. For example, the company's marketing reps use GIS to direct their marketing efforts toward those who are most likely to sign up for Qwest Wireless service. Subscribers can have one number for both their wireless and home phones, as well as a single voicemail box for their home, wireless phone, and office. This service also allows subscribers to use their phones to browse the Internet and check e-mail.

Because direct mailing is expensive, Qwest Wireless needs to limit the mailing list to the people most likely to sign up for service. So marketing reps query the GIS to find prospects who fit a target profile and can benefit most from Qwest Wireless service.

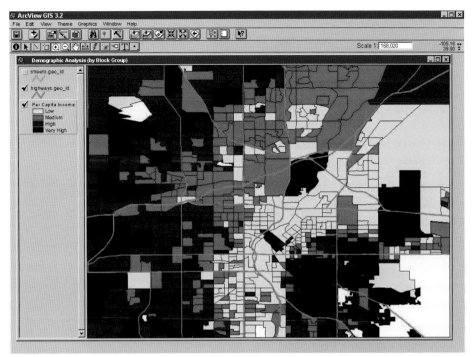

Maps like this one showing an area's population by income let marketing representatives see who would be most likely to subscribe to Qwest Wireless service.

Boosting sales

When customers call for information about Qwest Wireless service, hundreds of sales representatives are available to provide them with coverage information.

Qwest Wireless uses ArcView Internet Map Server to create coverage maps that the sales team can access with the GWIZ application. Sales representatives enter an address, intersection, or city to see if the caller lives within a Qwest Wireless coverage area. They can also use GWIZ to direct the customer to the nearest store where they can buy a phone and set up service.

Without GIS, these representatives would be looking at paper maps every time a customer called, and it would be impossible to keep all the paper maps current. GIS lets Qwest Wireless provide the current and accurate data that their customers expect and demand.

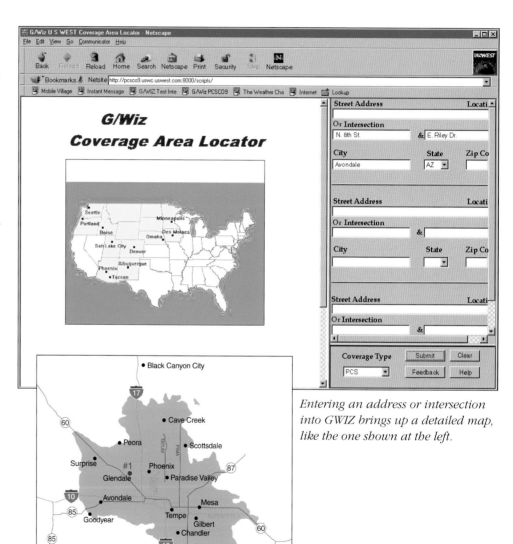

Entering an address or intersection into GWIZ brings up a detailed map, like the one shown at the left.

Planning ahead

In the future, Qwest Wireless plans to have customer service representatives input customer complaints directly into Wildcat. Being able to see where more customers are complaining will let analysts find weak spots in the network and make better decisions about which areas to fix first.

Qwest Wireless is also working to meet Federal Communications Commission (FCC) standards that, over the next few years, will require that all wireless carriers be able to accurately locate subscribers who call 911 in an emergency.

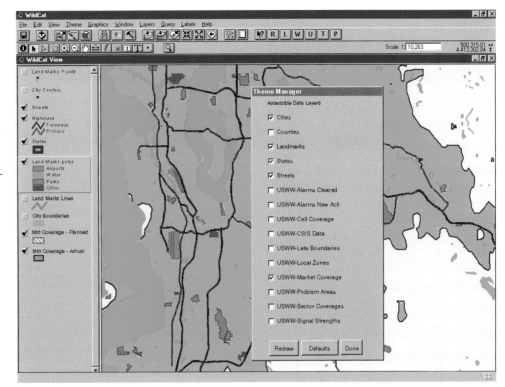

Currently, Wildcat's spatial database includes only information about Qwest Wireless' coverage areas. Soon, however, Qwest Wireless will expand Wildcat's spatial database to include information about where it carries roaming agreements.

The system

HP®-UNIX® workstations and Sun workstations running ArcView GIS, ArcView Spatial Analyst, ArcInfo, SDE, Oracle, and ArcView Internet Map Server

The data

Data layers in Wildcat include:

GDT landbase (streets, highways, landmarks, water features, cities, and county boundaries)

Qwest Wireless custom layers (coverage boundaries, signal strength boundaries, cell sector boundaries, cell boundaries, cell site base station locations, known problem areas, local access and transport area boundaries, and alarms)

The people

Special thanks to Mark Mullane, manager of GIS operations for Qwest Wireless.

5

Wireless, but still connected

Just a few years ago, wireless phones were as rare as cars were in 1910. Now it seems that everyone, from soccer moms to CEOs to school kids, has a wireless phone. Some people speculate that someday mobile phones will be our only phones.

The more people use wireless phones, the more intensely the people who operate mobile phone networks compete with each other. Given equal technological quality, the companies that compete the most successfully are the companies that provide the best customer service. In this chapter you will see how T-Mobil, a Deutsche Telekom AG subsidiary, is using GIS technology to integrate its spatial database management systems and to help its customer service representatives provide customers with quick and accurate answers to their service questions.

T-Mobil is part of Deutsche Telekom AG's rapidly expanding empire of telecommunications services. Even

though currently Europe's number one telecommunications carrier, Deutsche Telekom AG, along with its subsidiaries and affiliated companies, never stops looking for ways to improve its products and services.

T-Mobil's Service Centers, located throughout Germany, employ more

than four thousand customer service representatives who use a company-wide GIS called T-Map to answer customers' questions about everything from where T-Mobil offers service to where customers can expect to pay roaming charges.

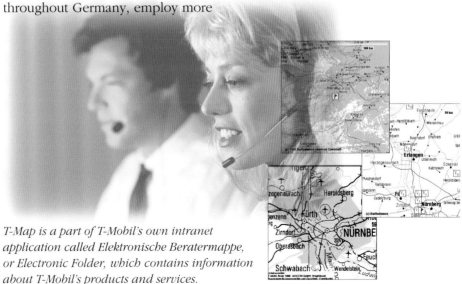

T-Map is a part of T-Mobil's own intranet application called Elektronische Beratermappe, or Electronic Folder, which contains information about T-Mobil's products and services.

Exploring all the options

Before T-Mobil developed T-Map, it stored service areas, tariff boundaries, and the like in the GIS on each call center desktop. This seemed like a good idea until it noticed that changes made on one desktop weren't always reflected on the others. As a result, representatives didn't always have the information they needed to answer customers' questions.

T-Mobil began to use advanced GIS server technology to integrate its spatial data into the company's intranet and extranet, as well as the Internet. This way employees throughout the company, as well as T-Mobil's external partners, dealers, and even customers, would be able to access and use the company's spatial information.

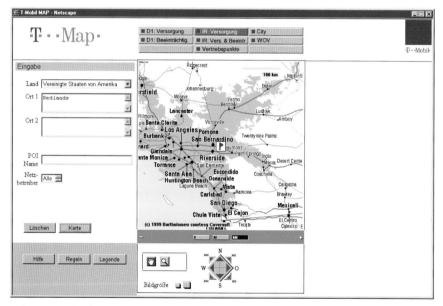

When T-Mobil finally introduced T-Map in its customer service centers, the system was so successful that the company immediately expanded it to its extranet and to the Internet.

Integrating the system

With its spatial data already stored in Oracle, T-Mobil looked for software that would let employees store, manage, and retrieve this spatial data quickly. ESRI's ArcSDE software was the answer.

While installing sophisticated software often means companies have to spend time and money modifying their current system, ArcSDE let T-Mobil merge its spatial data into its company-wide database without having to reconfigure its entire network or purchase all-new hardware. This way, T-Mobil was able to affordably build the enterprise GIS it was looking for.

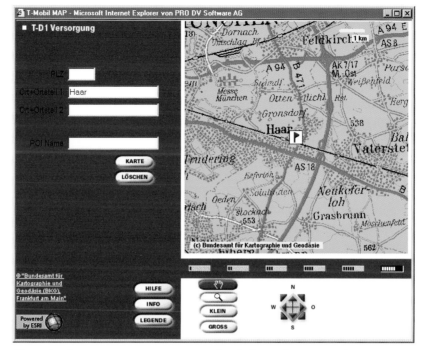

No matter how many users are requesting information at one time, T-Mobil's distributed architecture using ArcSDE lets users mix and match spatial data layers. T-Mobil can easily enhance T-Map by simply adding new topic layers whenever necessary.

Bringing it together

T-Mobil first had to copy all its spatial data into a single database. It used ESRI's ARC Macro Language (AML™) to combine the data sources from each workstation, storing this data as ArcSDE data layers in its Oracle DBMS. Using ArcSDE, it was then able to import and export data, moving data from one DBMS to another without corrupting, duplicating, or losing information.

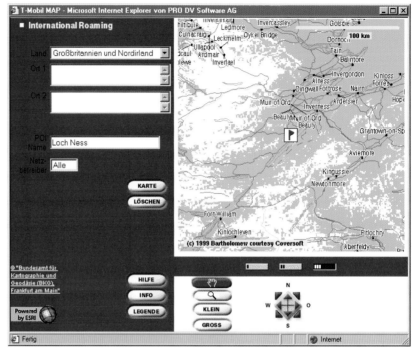

T-Mobil had a lot of things to consider as it set up its large-scale GIS, including how to collect new data, how to process this data, who its audience would be, who should have access to data, and how much access they should have.

Preventing errors

With data coming from so many different sources, T-Mobil needed a way to catch errors and duplicate information.

ArcSDE lets T-Mobil look for errors in point, line, and polygon information as it is processed. When creating a layer in ArcSDE, the user defines the feature type (point, line, or polygon) allowable for that layer. Lines are not allowed in a point layer; points are not allowed in a polygon layer; and so forth. Similar rules exist for polygons, which require, among other things, that a polygon's boundaries must be closed before the data will be accepted. When ArcSDE finds errors, users can fix them within the application.

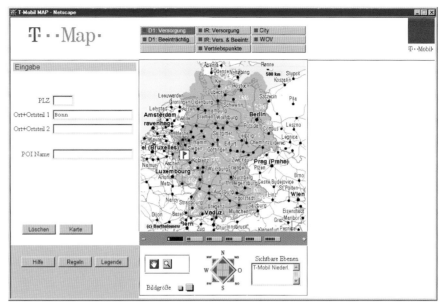

ArcSDE lets T-Mobil load and activate newly created data without having to shut down the server.

Making it uniform

Data that comes from different sources may use different resolutions, projections, or coordinate systems, each of which represent the data differently. Differences among some data layers often make it impossible to align the layers properly or even to view them together.

ArcSDE and ArcInfo allow T-Mobil to collect mapped information from different sources, in different coordinate systems, at different scales, or in various geographic projections, and make them consistent, either by converting them to the same coordinate system or by registering them to the same basemap. T-Mobil can then view these data layers together.

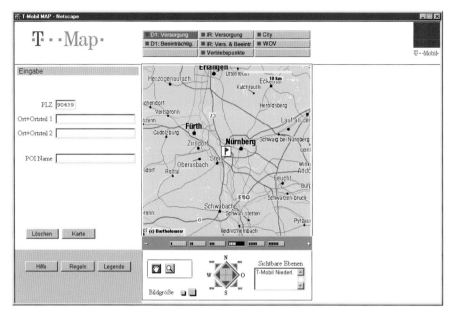

As it receives and processes each data set, T-Mobil checks to see if it is compatible with the rest of the spatial data. Whenever there are discrepancies, ArcInfo software is used to make the data uniform. The conversion algorithms used by ArcInfo are also applied to raster data as it comes in from outside vendors.

Accessing the data

T-Mobil decided how detailed the data needed to be, and who would have access to that data by identifying who would be using its GIS and viewing GIS maps. For example, marketing employees didn't need information about network failures, but those in the engineering department did. Inventory managers didn't need to know where dealers were located, but customer service representatives did.

Monitoring access to data was particularly important for T-Mobil because T-Map was set up so that people outside the company could access its spatial information. T-Mobil protects confidential data with firewalls and other security measures.

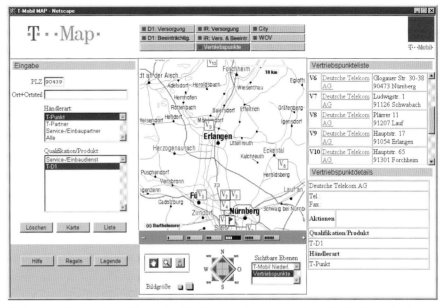

This map shows where Deutsche Telekom customers can buy T-Mobil products in the area in and around Nuremberg.

Up-to-date info, in seconds

By allowing any number of data layers to be viewed together, ArcSDE lets T-Map customer service representatives quickly retrieve data and create thematic maps. As themes are selected from a list in T-Map, ArcSDE works behind the scenes to retrieve the requested data from the DBMS instantly.

With customers waiting on the line, representatives had to be able to get a response from the online system within seconds of submitting a request. As each map is generated, it is updated with the most current data.

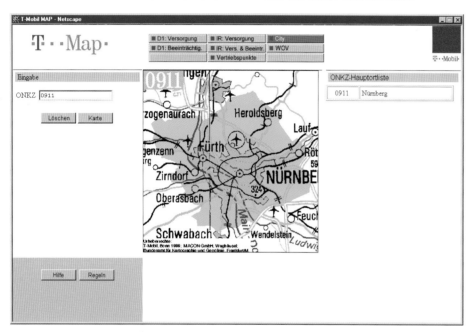

Since customer service representatives have to answer each customer's question as promptly as possible, the maps they view must be produced very quickly and be easy to read and understand, even by those with no GIS or cartographic experience.

Keeping it current

Spatial data that changes less frequently, like street data, will be updated less often. Other information, such as network failures, must be updated continuously, especially if these failures mean people will be without phone service.

To keep the data current within very short update cycles, T-Map uses a highly advanced algorithm. As data is requested, T-Map retrieves the most current data and returns a real-time map of the requested parts of T-Mobil's network. This assures a continuously up-to-date view of the mobile radio network.

The integration of spatial and non-spatial data lets T-Mobil provide better customer service, and saves time and money.

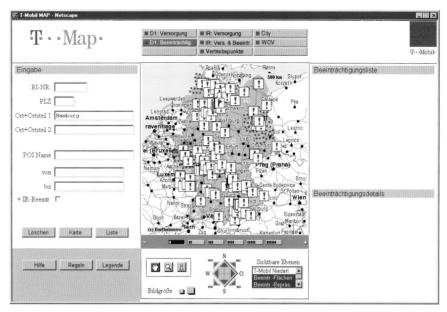

Business partners and customers who log on to T-Mobil's Internet GIS can sometimes answer their own questions, getting the information they need when they need it.

The system

In its customer care departments, T-Mobil uses Sun Enterprise 5000 servers, each with four microprocessors, 4 GB RAM, and 210-GB hard drives. It uses a Sun Enterprise 450 server with two microprocessors, 2 GB RAM, and a 210-GB hard drive as an Internet server, extranet server, and for data processing and distribution. Software includes ArcInfo, ArcSDE, and ArcIMS.

The data

T-Map's spatial data department provided vector data from Bartholomew, GSM coverage data from Coversoft, and data from the Bundesamt für Kartographie und Geodäsie.

The people

Thanks to Eugen Schulz, CIO of T-Mobil, and Manfred Kroll, senior head of the Department for Customer Care Information Systems; Karl–Heinz Recke, project manager for T-Map; and Mueed Haleem, project assistant for T-Map, all of T-Mobil.

T··Mobil···

Ancient antagonisms and thousands of miles of controlled borders have given way, in most of Europe, to monetary union and vigorously pursued commercial intercourse. The economic strength of the European countries depends on the ability to come and go, and to cooperate—an ability that in turn depends on effective communication. Travelers using mobile phones and moving from jurisdiction to jurisdiction, language to language, need to be able to stay connected.

It's no surprise that telecommunication companies throughout Europe, including those that provide mobile telephone services, are working toward becoming more and more accessible to more and more countries. In this chapter you will see how the various departments within KPN Mobile use GIS to manage their international digital cellular radio network, increasing accessibility and improving customer service while maintaining security.

KPN Mobile is a part of Koninklijke PPT Nederland (KPN), the oldest and most successful telecommunications company in the Netherlands. Formerly operated by the government, KPN has been a private company since 1989.

In July of 1994, KPN added GSM (Global System for Mobile communication) as an alternative to its analog mobile phone service. GSM is a digital cellular radio network that increases the amount of data that can be carried by dividing each cellular channel into three time slots. By year's end KPN had added 70,000 GSM customers to its 250,000 analog customers.

By the time the analog network closed in October of 1999, KPN already had nearly four million GSM mobile phone customers, making it the leading mobile phone company in the Netherlands.

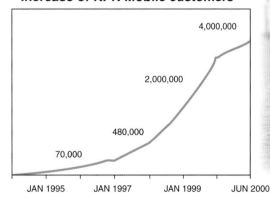

Increase of KPN Mobile customers

Adding digital technology netted KPN more than three million customers in less than six years.

Developing a global system

Before GSM technology was developed, each country in Europe had its own analog cellular telephone system—incompatible with everyone else's in equipment and operation. Not only could companies operate only within their own national boundaries, but with sales and services limited to subscribers within these geographic limits, each company was unable to manufacture and sell enough equipment to offer substantial savings. Until Europe could find a way to develop a system standard, industry growth there would remain quite limited.

GSM phones contain a card that identifies its owner's account to the network and provides authentication, ensuring accurate billing no matter which country the caller is in when making the call. This technology is now the standard for wireless phone service in Europe, offering nearly complete coverage in western Europe and growing coverage in the Americas, Asia, and elsewhere.

Spreading out

During the early 1990s, the European Telecommunication Standards Institute began publishing GSM specifications. Adopting these standards gives a company like KPN the chance to compete in a much larger market and to offer its customers more flexibility.

For example, GSM lets cellular subscribers use their wireless phones in any GSM service area in the world as long as their provider has a roaming agreement there. This means that the wireless phone a customer uses in the Netherlands could work in France, Germany, Australia, Finland, or even China.

Since GSM service is available all over the world, KPN now has to manage a much larger geographic area than before. With this tremendous increase in spatial data, KPN looked to GIS as a way to better manage its network.

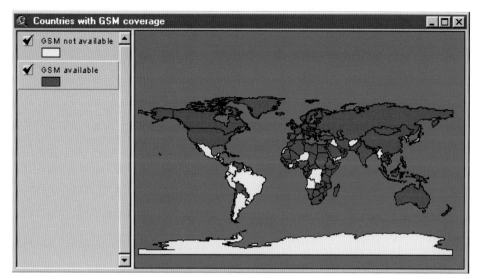

Created in ArcView GIS, this map shows countries where GSM coverage is currently available (red). More and more countries are adopting this technology to provide their citizens with mobile phone service no matter where they travel.

Introducing Octopus

When KPN first created Octopus, a custom application built with ArcView GIS, the company set out merely to design a GIS that could be used to manage its GSM coverages and roaming agreements. It ended up with much more.

The Information Services department manages how users from other departments enter data into the system and which system data those users can access. For example, as data comes in from other departments, the Information Services team processes it and arranges it in Octopus to ensure that it is compatible with the GIS and with company standards. This ensures that the data is free of errors and duplicate information, making it possible to integrate the different applications used by various departments. The team also set up firewalls to protect sensitive information like customers' names and addresses, so that each user can only see information that is necessary for their work.

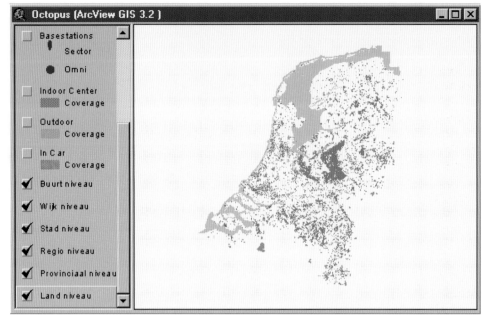

The name Octopus comes from the fact that there were originally eight users set up within KPN to use the GIS software. Octopus has been so successful that there are now 144 users throughout the company creating and viewing maps like this one.

Mapping coverage layers

Mobile phone customers sometimes complain about being unable to complete a call. When this happens, the complaints are registered in a Trouble Ticketing System and then plotted on a map in Octopus. By mapping where each customer was when they had trouble, analysts in the Service and Network Management Centre can easily see where the trouble spots are.

These analysts then map each customer's location with the different coverage layers of the GSM network. Examining the locations of the complaints next to information about KPN's GSM coverage in that area helps the analysts identify why the problem has occurred.

Another layer in the GIS contains measurements about the quality of KPN's network along roads and highways. KPN's cars drive through the Netherlands, using special equipment to measure where service is available. This information is later plotted on maps in Octopus to show the coverage along different routes where customers can make mobile phone calls and where there may be service interruptions.

Before Octopus was developed, none of the problems in the Trouble Ticketing System were mapped. Sometimes the same complaint was listed more

than once, often because more than one user entered data. Since there was no way to view the information graphically, it was difficult to find duplicate data. Sometimes several operators were busy handling the same complaint.

Implementing a GIS solved this problem. By using Octopus to generate maps, users are able to see where more than one complaint showed up as a single point, making large concentrations of customer complaints easier to identify.

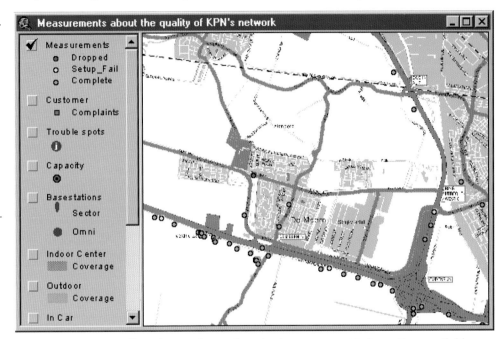

Measurements are collected every day so there is always current information available about the quality of KPN's GSM network.

Handling customer complaints

When several complaints appear in the same area, KPN's analysts begin to look closer at their network to determine the cause of the problem.

By comparing customer complaints with their own coverage data, they can often determine the cause of the complaint. Occasionally a customer will try to make a call in an area where service is still unavailable. More often, however, service was available, but the base station in that area was not large enough to handle the call volume.

Whenever the demand for service exceeds what the base stations can provide, the GSM network must expand. To keep costs down and still deliver high-quality service, companies like KPN have to expand their networks in the right places at the right times.

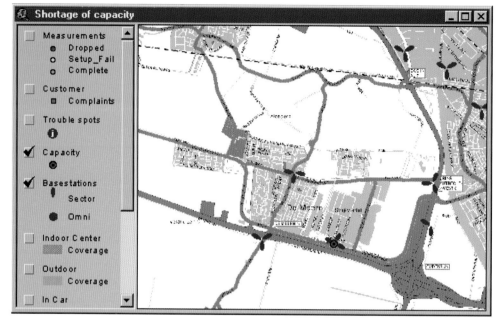

Each base station captures data about how much traffic it can handle, how much it is handling each day, and whether its capacity is enough. This information is imported into the GIS and plotted on a map.

Viewing events

By finding out ahead of time where demand is likely to increase, companies like KPN can sometimes avoid the problems that come from not having enough base stations or from base stations with insufficient capacity. To help KPN better anticipate these situations, the company's marketing department predicts mobile traffic growth using a GIS application called TIGER (Traffic Intelligence through Extrapolation Results), one of the many layers in the Octopus system.

Within TIGER, service areas are categorized according to which type of mobile phone usage is most prominent in each area. For example, calls may be grouped according to whether they occur in a car or not, or whether the call was made in a commercial or residential district. This makes it possible to relate mobile phone traffic to specific geographic areas, like urban areas, business districts, and highways. By combining the predicted number of mobile phone calls per area with information from marketing research or business plans, KPN can estimate how and where mobile traffic is likely to increase.

KPN then calculates the actual call volume in each area from results reported by each base station. Using a standard statistical package, analysts create a linear regression model from this data, which they then convert into a grid. The actual statistical analyses aren't performed in TIGER, but the results are input into the GIS so they can be mapped and then shared with other departments.

Through this process, the marketing department can not only predict growth in an area, it can also determine where larger capacities will be temporarily required. For example, the department uses GIS to plot the locations of all the large-scale, local events on a map in Octopus. These maps can then be used to determine whether the base stations in the area can handle the load. If not, KPN can set up more mobile base stations before the event.

When there is a large event, like a concert or football match, KPN employees can place a mobile base station in that area to increase service capacity during the event. They can also place a mobile base station near a base station that will be out of order for a while.

Planning for more

The Planning department uses these maps to help identify possible locations for base stations according to where traffic is and where it is likely to be. When surveying sites in a particular area, the planners must study each location suitable for building a base station.

When placing base stations, KPN's planners must consider whether their goal is to simply provide coverage or to increase capacity. In densely populated areas, base stations are often placed on or near buildings. When coverage is the primary concern, base stations are placed at the tops of the highest buildings, allowing the signal to travel farther. This does not, however, increase the number of calls the base station can handle. Now that coverage is available just about everywhere in Europe, capacity is more a concern. To increase the number of calls a network can handle, base stations are placed closer together and lower than before, concentrating their capacity and coverage in a more limited area.

Locations that match the planners' criteria are mapped. They then use the GIS to simulate coverage profiles for each. The locations with the best coverage and highest capacity in the required areas become the recommended sites of new base stations.

Without Octopus, hours and hours would be spent in the field, visiting locations and marking up paper maps. By using the power of GIS, all this work can be done with the click of a mouse.

KPN is working to implement three-dimensional mapping capabilities into its GIS so it can create maps like this one. Analysts will then be able to see the heights of buildings and view coverage density in three directions.

Offering instant solutions

Along with adding three-dimensional technology to its GIS, KPN plans to combine the Trouble Ticket system with Octopus, making it possible to plot a trouble ticket form instantly as a point on the map. This will let the Customer Care department immediately view information on a map while the customer waits on the other end of the telephone line.

Every department will thereafter be able to see the customer complaints on the map. So, when a customer calls with a complaint about an area where there have been other recent complaints, the operator will simply click on a complaint that is already in the system. When an operator clicks on a location, the GIS will display all the information about that customer complaint, including information about when a proposed solution will take effect.

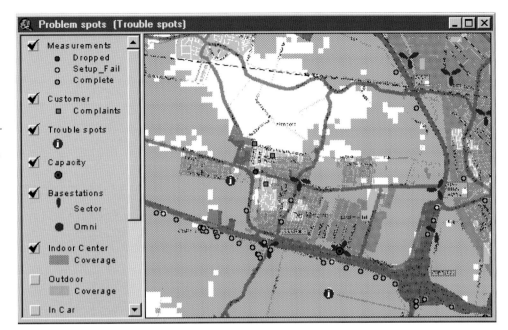

KPN's list of problem spots is also plotted in Octopus along with a projected solution for every problem. By referring to maps like this, the Customer Care department can give KPN's customers more informed answers, saving time because each inquiry does not have to be individually investigated.

The system

DEC™ (Compaq®) AlphaServer™ 2100 5/250, Compaq 1850R, running ArcView GIS and Octopus.

The data

Customer complaint data supplied by Mobiticket; coverage information by MSI Planet; postcode data by Bridgis, BV; demographic data by the Central Office for Statistics in the Netherlands; business data by the U.S. Chamber of Commerce; statistics about global GSM coverage by GSM World, Inc., and mapped in ArcView GIS by ESRI.

The people

Thanks to Gerard H. Kroon, manager, and to Frans Beukers, quality analyst, both of the Service and Network Management department at KPN Mobile.

Everything you need

7

You've reached a certain point in your life, or your career, and decide that a change is in order. You want to relocate, and you're tired of both the northern and the western hemispheres. Australia is the answer. You won't have to learn a new language, and the frontier is still wide open.

In this chapter you will see how Pacific Access Pty Ltd uses GIS technology to provide users with instant information about the location of just about any business, residence, or other point of interest. With myriad online mapping applications and services, users can also map neighborhoods or entire cities, e-mail a map, or even link a map to their own Web page.

Pacific Access, a subsidiary of Telstra Corporation, Australia's largest telecommunications company, displays more than two million maps each month from more than twenty Web sites. Using ESRI's ArcSDE software to store, manage, and retrieve spatial data from the company's Oracle database management system, Pacific Access provides many different interactive mapping applications to several Internet sites.

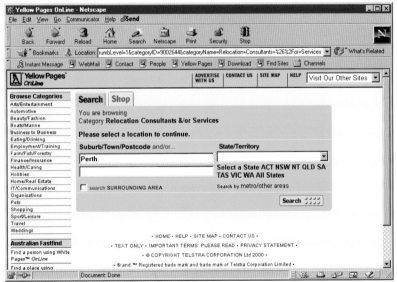

Links to the most visited topics, like finance, shopping, and real estate, are listed on the left side of this Pacific Access Web page.

Finding long-lost friends . . .

As you plan your move, you find you've misplaced the business cards of some contacts who work in Australia. And then you remember that an old friend from college planned to move there. You can recall his name, but that's it.

No matter who you're looking for, with Telstra's Australian White Pages OnLine, all you have to do is enter a name, state, and locality, and specify a residence or business. This way you could search all of Australia in just a few minutes.

Updated daily, the White Pages OnLine site receives more than 1.8 million visits per week. In addition to providing information like addresses and telephone numbers from all fifty-five Australian White Pages directories, the White Pages OnLine site uses Whereis Internet mapping technology to provide maps of locations users have searched.

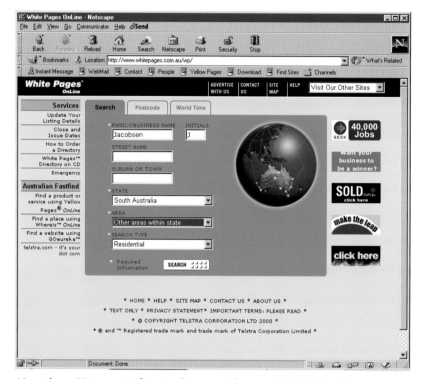

More than 59 percent of Australians say they use their White Pages directory at least once a week.

. . . and forgotten places

A move to the other side of the world is a vast and complicated undertaking; some things are bound to slip through the cracks, like the name and address of the restaurant in Perth where you had perhaps the greatest meal of your life, and where you'd like to celebrate your permanent return.

Telstra's Australian Yellow Pages OnLine lets you search an area by business name, category, or both.

Since you don't know the name of the restaurant, you simply type "restaurants" for the category, then enter the locality and state. Or you can click the Eating/Drinking link on the left under Browse Categories. You click on Dining/Eating out in the next list that comes up, then Restaurants in the following list. After you enter the locality and state, a list of restaurants appears, including the one you were looking for. A single click on the restaurant's name brings up the address and phone number, along with a link to a map of its location.

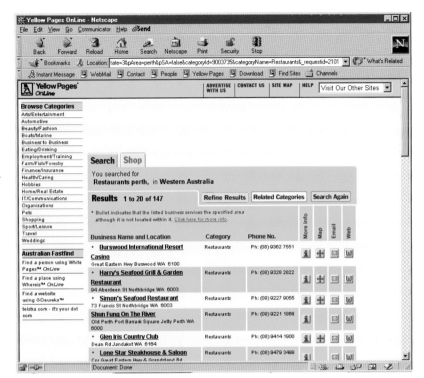

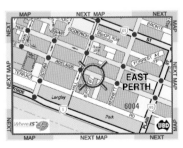

Most of the businesses on the Australian Yellow Pages OnLine have map links. Clicking on a map link brings up a map like the one at the left.

Where map links come from

Many businesses with their own Web sites, like that restaurant, have purchased map links from Telstra's Whereis Online site.

Clicking on the map link sends you directly to the Whereis Online site, where you can view a map of the business location.

These maps are created using Whereis Online, an online street directory of Australia's major cities, which uses maps from UBD™, a division of the Universal Press Pty Ltd, one of Australia's largest publishers of street directories and atlases.

The Whereis Online database includes nearly every street address or point of interest within the greater metropolitan regions of Melbourne, Sydney, Perth, Adelaide, Brisbane, Sunshine Coast, and Gold Coast, as well as many of the more rural areas around the country. Pacific Access continually updates the database and extends the coverage with more detailed and accurate information, working toward a comprehensive interactive atlas of Australia.

Following your dream

Until you're ready to start your own business, you'll need at least a part-time job to help pay the bills.

The Australian Yellow Pages OnLine lets you search for a job the way you search for a business or residence. Clicking the Employment link at the bottom of the screen lets you search a database of job postings by specifying parameters such as location, full or part time, industry, occupation, or keyword.

After browsing through the part-time positions near where you're moving, you find the perfect job. If you land this, you'll be able to earn money doing something you love.

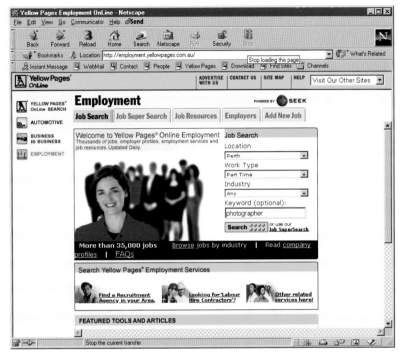

The Australian Yellow Pages OnLine lists thousands of employment opportunities in a comprehensive job search section. Once you find a position that suits you, you can even apply online.

You too can have a map link

Once you open your own business, you're going to need some employees of your own. You decide to add a map link to the Web site you're developing so potential employees, customers, and old friends can find you.

You start by creating a map on the Whereis Online site. First, you enter your business street address, including the suburb and state. The business location appears on the map with a cross-hair icon. You can pan and zoom on your map until the scale suits you, then recenter it if you want. Once you are happy with your map, you can e-mail it to people or buy a map link.

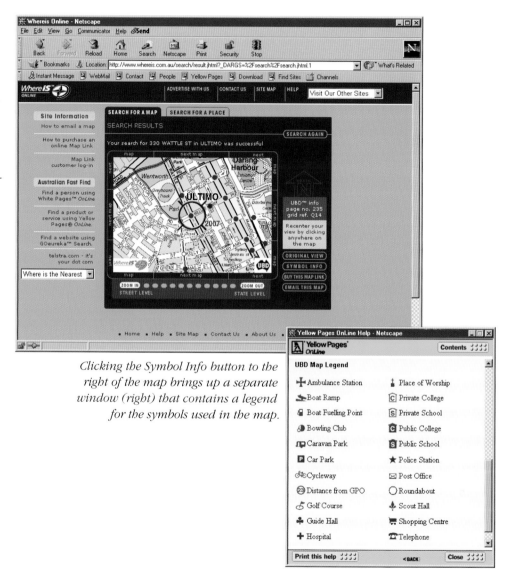

Clicking the Symbol Info button to the right of the map brings up a separate window (right) that contains a legend for the symbols used in the map.

Making it your own

Buying a map link means simply registering with the Whereis Online Service. By filling out the online customer registration forms, you create an account that you can use to manage your links. Each link will be valid for twelve months.

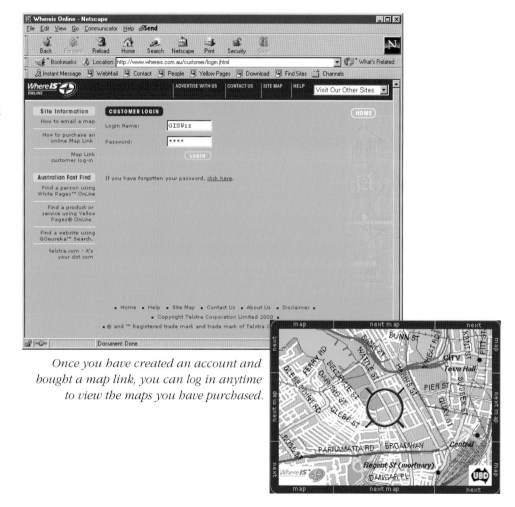

Once you have created an account and bought a map link, you can log in anytime to view the maps you have purchased.

Keeping it familiar

Next, you need to link your Web site to the Whereis Online site by adding the Whereis Map Link logo along with its associated code to your Web site. When visitors to your site click on the Map Link logo, they are sent immediately to your map on the Whereis Online site.

For businesses that have more than one location, such as bank branches, ATM machines, retail outlets, or government facilities, Telstra offers the Whereis Custom Connect™ service. The Custom Connect service lets these businesses add interactive mapping directly to their Web sites. With this service, users search maps within the business site rather than being redirected to the Whereis Online site, letting businesses control the design and content of their own mapping page.

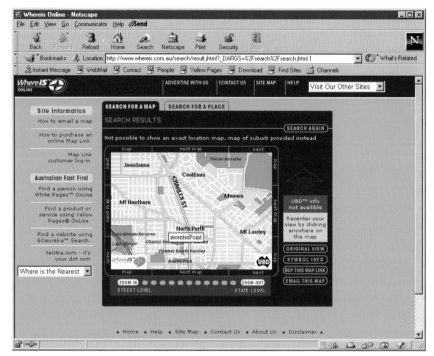

Whereis Custom Connect maps also include key maps, which show the map's location in relation to the rest of the city.

The system

Sun workstations running Solaris™ 2.5, Oracle 7.3.4, SDE 3.0.2, GIS Data Server 1.0, and a custom CGI interface built by Pacific Access.

The data

UBD data from Universal Press Pty Ltd and data from the Australian Land Information Group (AUSLIG).

The people

Thanks to Nick Gee, online GIS project manager for Pacific Access Pty Ltd, and to the Whereis Interactive Mapping Team at Pacific Access, for their contribution.

8

Getting around

Hong Kong's Victoria Harbor has for centuries brought the wealth and bustle of international trade to the city. The same harbor holds Hong Kong hostage against the spiraling mountains, hindering its ability to expand, and crowding every square inch of available land with residences and places of business.

Thousands of people come here every year to do business, thousands more simply to visit, crowding the streets even further. Whether you live and work in Hong Kong or are just visiting, finding your way around can be intimidating. In this chapter you will see how Cable & Wireless HKT uses GIS to maintain and develop a community map on the Internet so that both those who live in Hong Kong and those who visit can get where they want to go.

Cable & Wireless HKT (CWHKT), a member of the Cable & Wireless group, markets a wide variety of quality voice and data telecommunications services backed by a state-of-the-art, fully digital fiber-optic network. CWHKT offers the people of Hong Kong a variety of ways to communicate, including basic telephone service; Interactive TV (iTV); international calls; Internet access; mobile phone service; multimedia services; and satellite links.

CWHKT manages more than 3.6 million telephone lines, more than one line for every two people in Hong Kong.

Competing for business

The exclusive right CWHKT had to provide local telephone services expired in mid-1995, resulting in competition from three new local phone companies.

This helped spur CWHKT to take advantage of GIS technology both in managing its landbase data and in making it available to the public.

Today, CWHKT has one of the largest and most sophisticated GIS applications in Hong Kong, including its own AM/FM/GIS (automated mapping, facilities management, and geographic information system). Its AM/FM/GIS has a total of 150 workstations and fourteen symmetric multiprocessing (SMP) servers, which work to balance the workload among computers, allowing the system to serve more users faster.

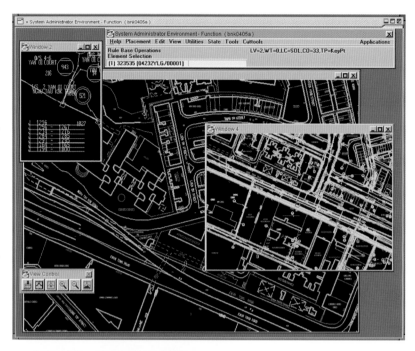

Engineers at CWHKT use GIS to manage more than 1.5 million objects, most of which are outside plant facilities, like underground cables, manholes, and ducts.

Mapping the city

In April 1999, CWHKT joined with the Hong Kong Lands Department and with Telecom Directories Limited (TDL), itself a joint venture of CWHKT and Bell Actimedia of Canada, to put a community map on the Internet. This mapping application was designed to allow people access to information on the Internet about dining, shopping, and transportation in Hong Kong, no matter which device they happened to be using.

With digital mapping data from the Lands Department, CWHKT and TDL used the latest database and Web technologies to create the Hong Kong City Map. People from all over the world visit this Web site; many live and work in Hong Kong, others are planning to visit. Perhaps even one of these is you.

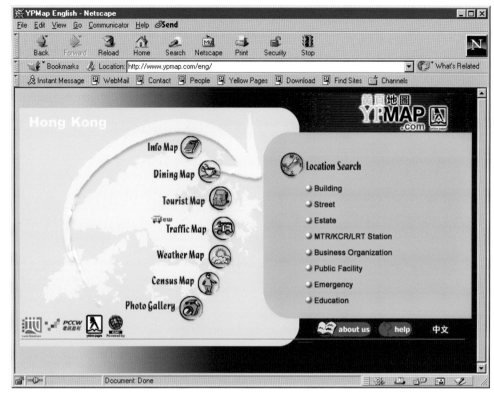

Even though Britain no longer rules Hong Kong, maintaining a high standard of English is still one of the city's greatest assets as an international center for business, finance, and tourism. For this reason, most businesses, including the Hong Kong City Map, conduct business in both English and Chinese.

Looking for a place to stay

Since you will be traveling to Hong Kong for an upcoming convention at the Hong Kong Convention and Exhibition Centre, you decide to use the Hong Kong City Map to plan your trip. Before you can look for a hotel, you need to locate the convention center itself.

You start by clicking on Building in the list of search options at the right. Because you don't know which district the convention center is in, you simply type "Hong Kong Convention" in the Building Name field, then click Search. Click on the link provided by the search, and the map automatically zooms in on the convention center's location, outlining it in red.

Once you've mapped the convention center, you click the Tourist Map tab, then select Hotels to find the ones near the convention center.

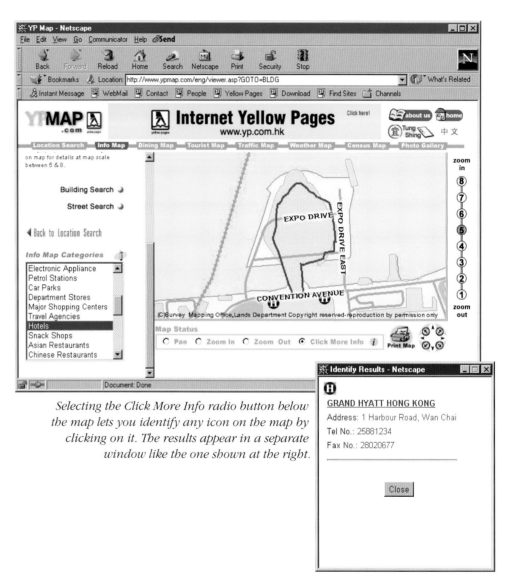

Selecting the Click More Info radio button below the map lets you identify any icon on the map by clicking on it. The results appear in a separate window like the one shown at the right.

Searching for food

You decide on a hotel right next door to the convention center. To recenter the map onto this hotel, you click the Tourist Map tab, choose Hotels, then type the name of the hotel into the Hotel Name field. Click on the name in the list that appears below. The map redraws with the hotel highlighted in red.

Although the hotel will undoubtedly serve great food, you'd like to find out what else is available within walking distance. Choosing Restaurant in the list at the left displays an icon for each restaurant in the area. Clicking an icon lets you identify the restaurant.

Once you select a hotel, you can scroll down the list at the left to search for nearby hotels, shopping centers, or restaurants.

Finding more options

You could also have returned to the home page and clicked on the Dining Map tab. The Dining Map allows you to search for a restaurant by food type, district, or both. From here you can also click on a link to the HK Dining Guide which allows you to search through more than seven thousand local restaurants. Occasionally, hyperlinks to restaurants' individual Web sites are available, letting you view menus and promotional offers. Both the Dining Map tab and the HK Dining Guide let you view the results of your search on a map.

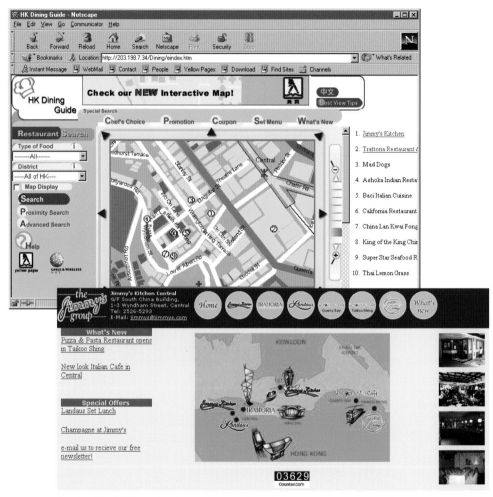

In the HK Dining Guide, each numbered restaurant corresponds to a restaurant on the list at the right. Restaurant names that appear highlighted have links to the restaurant's Web site.

Checking out the hot spots

Although the convention only lasts for three days, you will be in Hong Kong through the weekend and will have time to see some of the sights. You aren't sure where to begin, but luckily, CWHKT has made it easy to locate popular attractions all over town.

First you click on the Tourist Map tab, then Scenic Points. On the map that appears, you can either click on the name of a district from the list at the left, which displays general information about that district, or you can click on a lettered point on the map, which identifies that tourist spot.

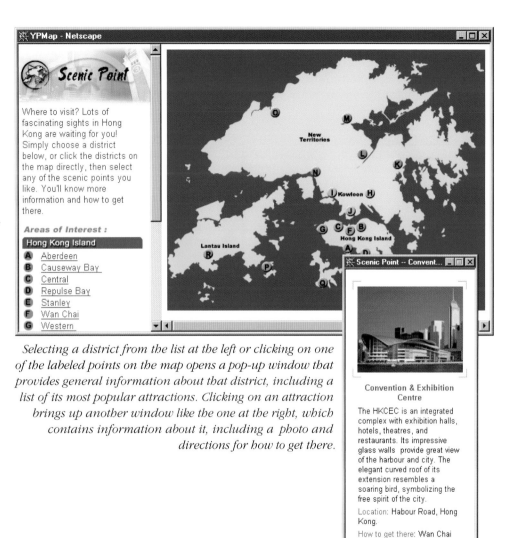

Selecting a district from the list at the left or clicking on one of the labeled points on the map opens a pop-up window that provides general information about that district, including a list of its most popular attractions. Clicking on an attraction brings up another window like the one at the right, which contains information about it, including a photo and directions for how to get there.

Seeing it all

You noticed on the Scenic Points map that there were tourist spots scattered all over the map. Deciding how to see as many of them as possible in the shortest amount of time seems overwhelming.

The Virtual Tour option offers suggestions about which sights to see, which order to see them in, and how best to travel from one spot to the next.

Once you select a district from the drop-down list at the top of the window, you can follow the virtual tour throughout the district by clicking on the time slots or on the numbered locations on the map.

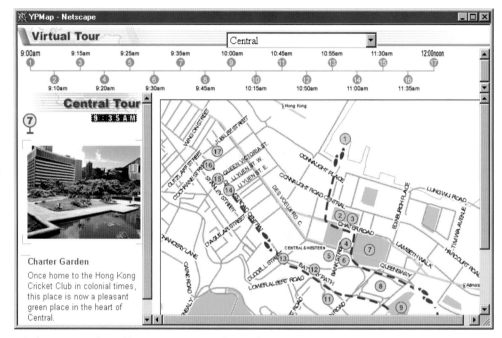

Click on a numbered location to see a photo of that sight along with some interesting historical facts about it.

Change is good

As the fundamental mapping information platform for Hong Kong, the Hong Kong City Map is also part of CWHKT's goal of building a comprehensive online Hong Kong city guide, aimed at creating a virtual Hong Kong on the Internet, one that changes as Hong Kong changes.

The system

Dell® 6400 Enterprise Servers running Chinese Microsoft Windows NT® 4.0, Oracle 8.0.4, Microsoft IIS 4.0, ArcIMS 3, ArcSDE 3, ArcInfo 8.

The data

Data from the Hong Kong Lands Department, the Hong Kong Yellow Pages, the Hong Kong Tourist Association (HKTA), and the Hong Kong Census and Statistics Department.

The people

Thanks to the GIS department at Cable & Wireless HKT.

CABLE & WIRELESS
H K T

9

Changing with the times

Each new telecommunication service— pagers, wireless phones, Internet access— seems to create a need for more services, yesterday's plethora turning into tomorrow's dearth. Telecommunication companies scramble to keep up with these insatiable needs, adding more to their networks every day in an effort to be all things to all people.

With each network handling more traffic than ever before, these companies look for new ways to manage their telephone or computer network. In this chapter, you will see how GIS software created by Telcordia Technologies, Inc., helps telecommunications companies keep up the pace.

To meet their customers' needs, Telcordia and its subsidiary, MESA Solutions, developed Telcordia™ Network Engineer, an application for designing and documenting fiber, copper, and coaxial networks. By basing this software on ESRI's ArcInfo and ArcSDE,

Telcordia was able to combine the best in GIS technology with a comprehensive set of network engineering tools specific to telecommunications. As a result, Network Engineer lets network operators design and document their telecommunication networks using not only points and lines, but also the components of the network such as cables, splices, structures, and cabinets.

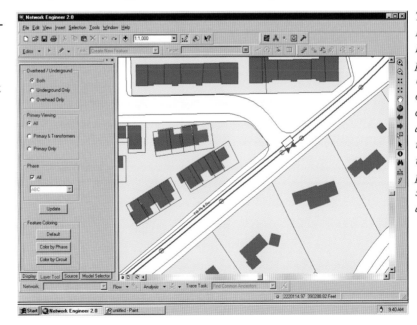

Telcordia Network Engineer provides layer-by-layer views of outside copper, fiber, and coaxial networks, as well as inside plant floor space, racks, and plug-ins.

Modeling the real world

Modeling things like power lines involves little more than moving from point to point. But telecommunication systems, like those that use fiber-optic cable, are more complex than power lines. Not only does each fiber-optic cable have attributes—owned or leased, aerial or buried—but each of the two to two hundred fiber-optic strands within each cable also has its own attributes. Some companies design, manage, and even lease strands of fiber within their networks. This complexity requires specialized GIS software for accurate modeling of the network.

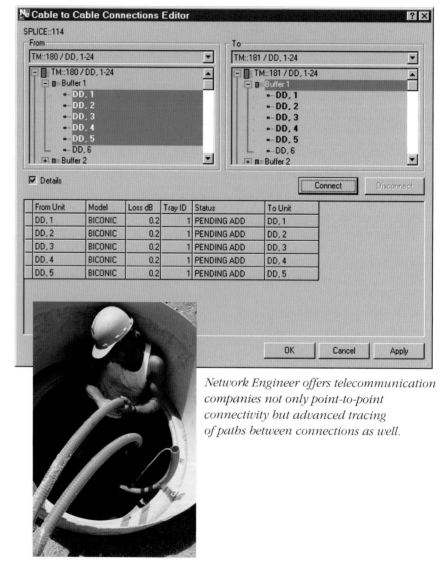

Network Engineer offers telecommunication companies not only point-to-point connectivity but advanced tracing of paths between connections as well.

Organizing the information

This need for a specialized GIS prompted Telcordia to work with ESRI when creating Network Engineer. Telcordia built Network Engineer around ESRI's ArcInfo 8 object-based data model, the geodatabase, which allows attribute information for objects to be combined with information about how those objects are connected within the network. Using the geodatabase, designers can do more than simply add or delete objects like structures, cables, and other equipment from their model. Now, they can associate rules about how objects connect and interact with one another.

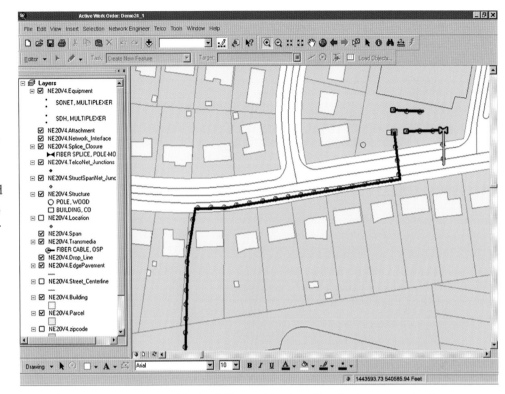

Network Engineer, working with ArcSDE, provides data for all telecommunications activities. Working as both an engineering application and a GIS, Network Engineer eliminates the confusion, inaccuracies, and duplication that sometimes plague a traditional communications network management system.

Defining the rules

Network Engineer comes with a geo-database that includes geographic objects commonly associated with a telecommunication network; a model-building module that provides a way to customize facilities or create new facilities; and a set of rules about how those objects may be connected. Rules associated with geographic objects like poles, manholes, and ducts prohibit the objects from being associated with other objects unless certain association rules exist. For example, Network Engineer will not allow a fiber cable to be connected to a copper cable without the right element between them. It won't allow a cable to be placed inside a duct if it's already connected to a pole.

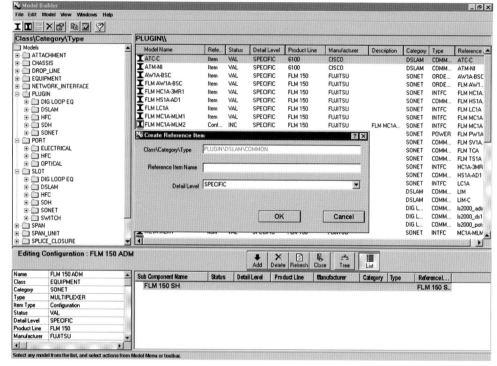

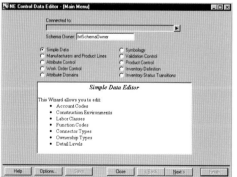

Using the Network Engineer Model Builder and the Network Engineer Control Data Editor, network operators can customize the rule base by adding new connectivity rules or features, or by deleting those that don't apply to their network. This allows engineers to create computerized models that match their particular networks.

Maintaining accuracy

As engineers post changes to the network, the system checks to see if changes follow the rules, and it prompts them to correct any errors. The hundreds of connectivity rules programmed into Network Engineer make the network operator's job easier by preventing users from entering erroneous data into the database. Since network operators can rely on its accuracy, managing their database with Network Engineer means they spend less time manually verifying the accuracy of their plant records.

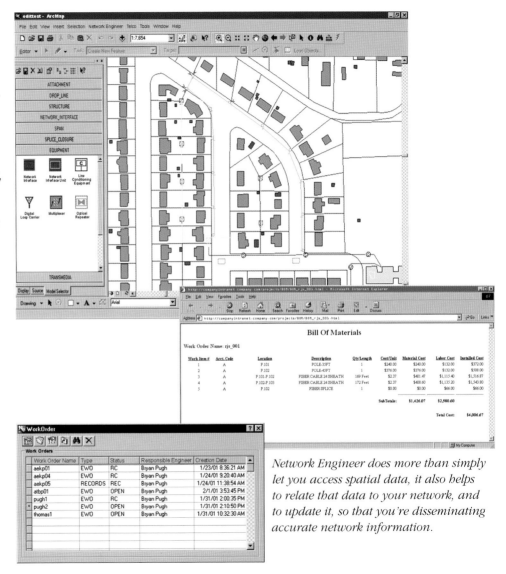

Network Engineer does more than simply let you access spatial data, it also helps to relate that data to your network, and to update it, so that you're disseminating accurate network information.

Inventory control

The changes engineers make to the network affect information inside the plant. Mapping the space inside the plant, including details of all storage racks and equipment, makes it easy to keep track of inventory. As items are used, Network Engineer records those items so they can be used to create a bill of materials.

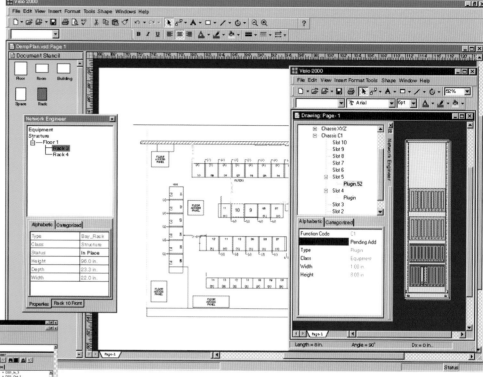

Unique, customizable symbols represent the status and the ownership of network inventory items.

Setting the standard

As they are completed, changes to the network are updated in the company-wide database. Employees from engineering to marketing can use Network Engineer to execute spatial queries on specific geographic areas, such as locating poles in a particular neighborhood, mapping customers in a city or state, or identifying the housing developments expected to be completed in a particular area within the next five years.

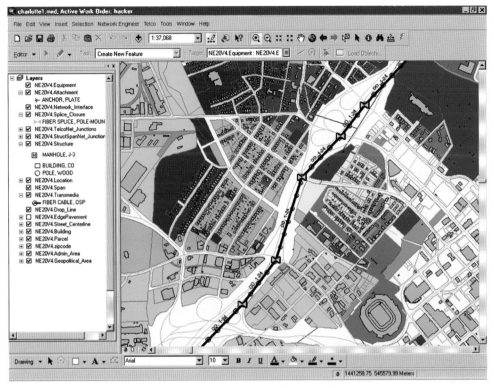

Every telecommunication company needs a comprehensive data management system to keep track of the rapid changes in its network, and to provide good service to its customers.

Getting started

Network Engineer is easy to use, because the drag-and-drop, point-and-click interface of ArcInfo 8 makes accomplishing complex tasks as easy as moving a mouse.

To gain familiarity with Network Engineer telecommunication tools built on ESRI's ArcInfo 8, the consulting and systems integration group at MESA Solutions offers a Network Engineer quick-start training program. This program is available at the client site or at the MESA training facilities in Huntsville, Alabama. Both Telcordia and MESA Solutions recommend the training class as the easiest way to gain an understanding of how the telecommunication tools of Network Engineer work in conjunction with ESRI technology.

MESA Solutions offers consulting services for those who purchase Network Engineer, including data conversion, custom training, product customization, and integration with Telcordia and third-party software applications and databases.

The system

ArcSDE, ArcInfo 8, and Oracle, running on UNIX, or Windows NT servers running Microsoft server technology.

The data

Any data readable by ArcInfo can be incorporated into the network model.

The people

Thanks to the entire Network Engineer product management, marketing, and sales team at Telcordia Technologies and MESA Solutions.

10

Gaining ground

For some companies, the Internet super-highway sometimes seems more like a dirt road. Slow dial-up connections. Laggardly downloads. Dropped connections. So much for the fast lane. But with more and more new businesses opening every day, fast, reliable Internet connections and high-quality local and long-distance phone services are a must, especially since making it big means competing with much larger companies, many of which can afford expensive fiber-optic connections.

In this chapter you will see how Teligent®, a global leader in broadband communications, uses GIS to help build its SmartWave™ broadband fixed-wireless network, offering businesses around the globe an affordable way to enjoy the benefits of broadband communications services.

Teligent first launched service in 1998 in Vienna, Virginia. Since then the company has grown to serve major markets throughout the United States.

By currently expanding its global presence through partnerships in Germany, France, Spain, Hong Kong, and Argentina, Teligent continues to work toward extending its reach into Europe, Asia, and South America.

Teligent's SmartWave technology offers multiple, fast Internet connections

and cheaper and more reliable local and long-distance calling. Teligent also offers quick and secure data transfer, as well as many other services, using digital microwave communications to send voice and data over very high radio frequencies at fiber-optic speeds.

Teligent's SmartWave network offers a solution to the so-called last-mile technology problem of bringing high-bandwidth services to users at an affordable price.

Sending signals

Fewer than 5 percent of commercial buildings are served by underground fiber optics. Fixed wireless radio technology called local multipoint distribution systems (LMDS) enables Teligent to supply customers with big bandwidth whether the buildings they're in have fiber networks or not.

Teligent delivers fixed wireless services by installing small antennas on the roofs of buildings. When a customer makes a telephone call or accesses the Internet, the voice, data, or video signals travel over the building's internal wiring to the rooftop antenna. These signals are then digitized and transmitted to a base station antenna on another building, usually less than three miles away.

Each base station antenna gathers signals from a cluster of surrounding customer buildings, aggregates the signals, and then routes them to a broadband switching center. The traffic is then distributed to other networks, such as public circuit-switched voice networks, the packet-switched Internet, and private data networks.

Today thousands of buildings are connected to Teligent's SmartWave communications network with antennas like the one shown here.

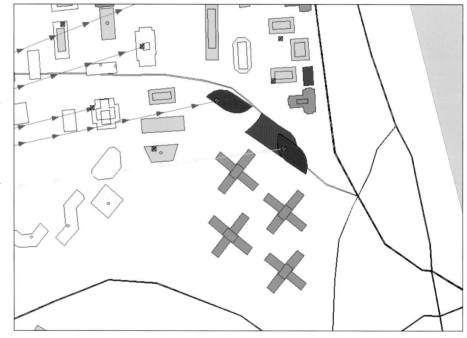

Mapping radio waves

To begin designing an effective network, Teligent must first locate unobstructed lines of sight between the transmitting and receiving antennas mounted on commercial buildings.

A line of sight is a three-dimensional concept, linking a tall building at some location with a shorter building at another location. In order to model lines of sight, engineers need to account for the difference in elevation, along with the difference in latitude and longitude.

Moreover, it isn't enough to climb to the top of building A and stare at building B through a pair of binoculars. Radio waves behave differently from light waves, so no matter how sophisticated your binoculars may be, what might look like a clear line of sight may not actually be one.

As important as line of sight is the area around a radio signal known as the Fresnel zone. An object in the Fresnel zone between the transmitter and the receiver can have a deleterious effect on the radio signal's strength and stability, even if that object isn't directly blocking the line of sight. The size of a Fresnel zone can be calculated based on the frequency and length of the signal itself; the nature of the obstruction itself is also a factor. Modeling Fresnel zones lets radio frequency engineers see whether a signal will be able to pass by obstructions without losing strength.

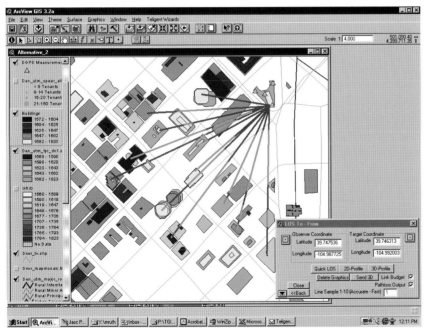

Since a large obstruction can reduce or totally block a radio frequency signal, engineers look for lines of sight with fewer obstructions. In this graphic, green lines show clear lines of sight, while red lines indicate likely obstructions.

Making the most of it all

With so many buildings and possible antenna locations, it would require endless hours and extensive resources to evaluate every alternative with a field visit. Using GIS, radio frequency engineers can simulate the path that radio waves and Fresnel zones will take from a particular location. Engineers then modify the location and properties of an antenna in the GIS to see how various changes might affect the signal propagation. Sometimes even a small relocation of an antenna site from one side of a rooftop to another, for example, may produce a substantial improvement in the radio network design.

Radio frequency engineers must account for various obstructions as well as other radio signals in the area that may interfere with or intersect transmissions and so reduce network reliability. Although obstructions caused by buildings or trees are often avoided, some obstructions may actually help the engineer by blocking out radio waves that would otherwise interfere. Also, by carefully mapping the propagation of radio waves, engineers may even be able to reuse part of the spectrum being used to serve another building. Reusing the limited amount of frequency allowed may help to fit more signals into a dense area and to serve more buildings.

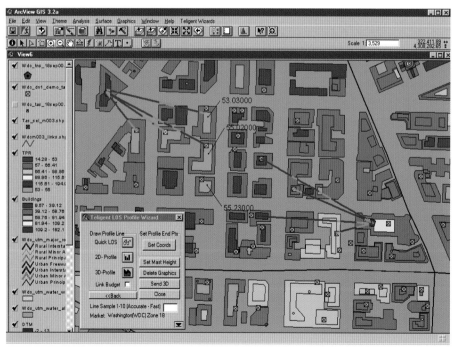

Some rooftops (light blue in this ArcView GIS view) are high enough to provide blockage between nodal service areas, making it possible for Teligent's engineers to reuse frequencies.

From footprints to top-prints

Radio frequency engineers use three-dimensional geographic data to decide where to propose and place antenna sites. This "3D Geodata" comes from mapping themes based on aerial photography, digital terrain models, and reflected surface elevation models showing both built and vegetated surfaces. Orthorectified aerial photographic images are frequently used, as is data about streets, business locations, and geographic market intelligence.

The aerial photographs are scanned, corrected, and loaded into a digital photogrammetry workstation, where they are used to construct three-dimensional views of the rooftops and of the precise location of rooftop features such as equipment sheds, elevator and stairwell housings, and parapets—places to mount Teligent antennas. The latitude, longitude, and elevation of these rooftop features, which are referred to as top-prints, can be measured in the GIS to an absolute accuracy of 2.5 meters.

Several data layers are created from the photographic images. These include polygons that outline the outer edge of the rooftop, also known as the footprint of the building, as well as the polygons that describe the three-dimensional locations of the rooftop features, or top-prints. This spatial data represents the various mapping surfaces required for modeling radio waves, and is collectively referred to as 3D Geodata.

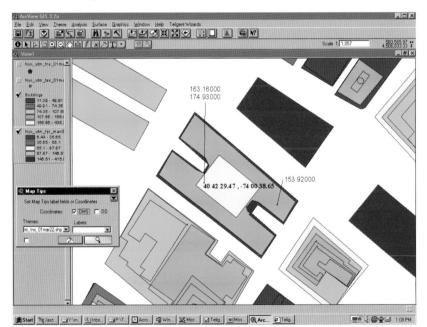

The digital maps engineers examine contain outlines of the rooftops and rooftop features like sheds and elevator and stairwell housings for all the significant buildings in each city. The rooftops are mapped to an accuracy of 2.5 meters (10 feet) in longitude and latitude (x,y) and elevation from mean sea level (z), and delivered as an integrated set called 3D Geodata.

Tools of the trade

When a business requests Teligent's voice or data services, field technicians go to the building to see if it's a good candidate for an antenna. They use SDE and Oracle to execute queries against the Teligent database to find out which buildings are already "lit," leased, or planned. They use differential GPS (DGPS) to collect geographic measurements like latitude, longitude, and elevation. And they evaluate the in-building telephone infrastructure, the rooftop power supplies, and the structural capacity of the building to support the proposed Teligent antennas and related equipment.

Using the Teligent Field Data Viewer, a tool created with ESRI's MapObjects® LT, field technicians compare Teligent's 3D Geodata with the data they collected using DGPS. The technicians highlight any discrepancies, then pass this data on to the engineers.

Field technicians on the rooftop can compare their differential GPS measurements (left) to Teligent's 3D Geodata layers using the Field Data Viewer (below). This map shows building footprints in blue, top-prints in pink, and an elevation measurement flagged in the center of the screen.

Easy to learn, easy to use

Teligent's engineers use ESRI's ArcView and several of its extensions, including ArcView 3D Analyst™ and ArcView Spatial Analyst, to analyze and model radio waves according to the 3D Geodata and the data collected and compared by the field technicians. Each of the extensions allows engineers to perform a different type of analysis.

Teligent used ESRI's Avenue™ programming language to create a set of specialized software wizards that run in the ArcView environment—adding menus, data entry forms, map views, layouts, and reports specific to radio frequency engineering. Adding the technical terms of radio engineering to the ArcView environment makes it easier for engineers to combine data about how radio waves travel with the spatial information contained in the 3D Geodata layers.

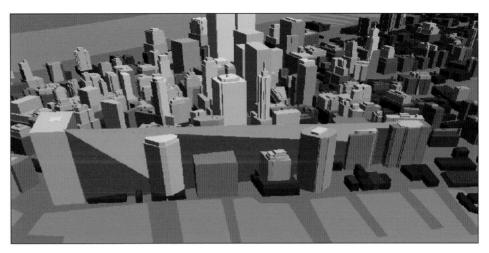

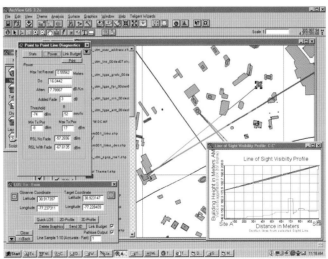

Specialized software wizards make it easy for Teligent's engineers to use ArcView and its extensions, including ArcView 3D Analyst, which they use to create three-dimensional maps of the paths of radio waves.

Speeding things up with spatial analysis

Engineers use ESRI's ArcView Spatial Analyst extension to create a viewshed to see all the potential lines of sight from a proposed antenna location. A viewshed shows which positions within a region are visible from a specified viewpoint. ArcView Spatial Analyst calculates a viewshed from a digital model of the land and rooftop surfaces that incorporates elevation as well as x and y locational data.

Digital terrain models are raster images, which means that the data they contain is represented as a series of cells in a grid. ArcView Spatial Analyst analyzes the slopes and elevations of all cells between an antenna and the cells within the selected area to determine whether a cell is visible from the antenna location.

Radio frequency engineers can use ArcView Spatial Analyst to model a seemingly infinite number of alternatives for antenna placement, varying the antenna location and mast height as needed to optimize the lines of sight to a customer's building.

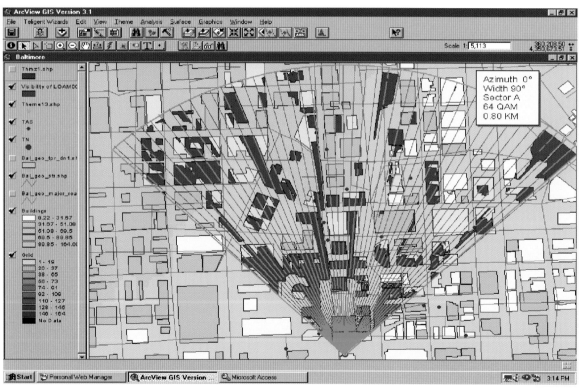

To produce a viewshed, engineers select a potential antenna location, an elevation above the surface, and a sweep angle.

Sharing the view

Once engineers determine the proper locations for new antennas, they often print a large map of those locations using ESRI's ArcPress™. These maps show Teligent's Nodal Area, all the buildings, antenna locations, lines of sight, lease status, and any other related telecommunications infrastructure and information. Without ArcPress, it would be impossible to include grids, images, or other raster data types, all of which are common to radio frequency engineering, in the plots.

Teligent plans to continue to develop its use of GIS, exploring ways to use three-dimensional geodata in marketing, sales, and leasing activities, and to contribute to the entire life cycle of business, from initial market planning through design, implementation, and network operations.

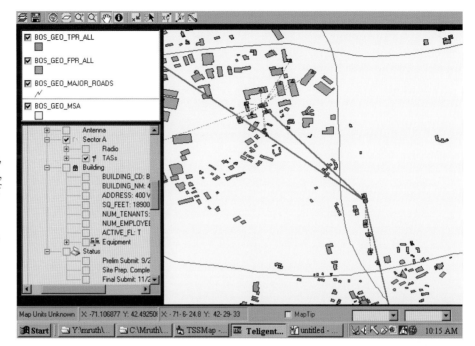

Teligent's mapping tools let members from the sales, leasing, operations, engineering, and management teams share a common view of the network. In the design review (pictured at right), business stakeholders view a MapObjects projection of the mapping database and make decisions based on the same information.

The system

Dell Optiplex GX1P running ArcView GIS, ArcView 3D Analyst, Avenue, ArcView Spatial Analyst, SDE, Oracle, MapObjects

The data

Maps are based on 3D Geodata provided to Teligent specifications by i-cubed, LLC, of Fort Collins, Colorado.

ESRI ArcData^{SM}, National Oceanic and Atmospheric Administration, and U.S. Geological Survey data sets were also used.

The people

Thanks to Mike Ruth, Tita Thompson, and Jubal Harpster of Teligent.

11

Bridging the gap

Shortly after Samuel Morse invented the telegraph, the streets of America and Europe were strewn with poles and lines carrying messages as fast as people could send them. But messages on both sides stopped short once they reached the Atlantic, a two-thousand-mile gap in what would otherwise have been the world's first international network. But by 1858, the pioneers of telecommunication had installed the first successful transatlantic submarine cable between England and Canada. In the short period between then and now, thousands of cables have been laid along the floors of all the world's oceans.

Today, the Internet explosion, along with the deregulation of the world's telecommunication markets, continues to increase the demand for communication bandwidth. But the large number of submarine cables already on the world's seabeds has made finding corridors and landings for new cables difficult. Helping resolve this bandwidth bottleneck is Concert, a joint venture of AT&T and British Telecommunications. Concert provides global communications services using fiberoptic connections laid under the ocean and along waterways from continent to continent and from city to city. In this chapter you will see how Concert uses GIS to help manage its fiber-optic submarine cable network.

Established as an independent company in 1999, Concert provides a variety of services designed to let multinational companies, traditional and emerging carriers, wholesalers, and Internet service providers develop global networks.

To reflect its global activities, Concert has offices in forty-nine countries around the world, including Australia, Bermuda, France, Germany, China, Italy, Japan, the Netherlands, Singapore, the United Kingdom, and throughout the Americas.

Today's submarine cable networks are larger, more expensive, and provide more bandwidth than ever before.

Call before you trawl

Located in Morristown, New Jersey, Concert's Undersea Systems Engineering and Records group maintains cable system data for hundreds of submarine cable systems. This group is responsible for the production, distribution, and maintenance of the Concert Submarine Cable Records for the systems where Concert is the maintenance authority. The group also supports the International Systems Maintenance Department, representing multiple system owners on maintenance issues, supporting the Cable Protection function for existing submarine cables, planning where to lay new cables, and supporting the shipboard teams that repair them. The records they keep include survey, installation, fault, and repair data; the latitude, longitude, and water depth at various points along the cable route; and attributes such as the type of cable or cable hardware. This information is essential for long-term cable protection and maintenance.

The Undersea Systems Engineering and Records group updates its database regularly, whenever existing systems have been repaired, data about newly installed systems becomes available, older systems are recovered, or new systems are planned.

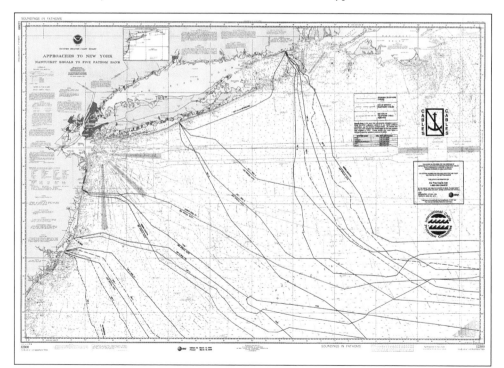

By providing both commercial and recreational boaters with maps that show where its cables are located, Concert helps prevent damage to these cables from fishing trawls and anchors.

The missing ink

In the early days of submarine telephone cable systems, all of Concert's submarine cable records were recorded and stored in an extensive collection of hand-recorded paper logs, typewritten reports, and Mylar® charts. This system was very effective in its day and the majority of these files still exist today. With the advent of fiber-optic submarine cables in the mid-1980s, and with Concert's push to deploy a worldwide fiber-optic network, these pen-and-ink records were computerized. All this historical data was loaded into a proprietary UNIX-based system that was the predecessor to Concert's new Cable Analyst GIS.

In the past, whenever a ship laid a cable, the Records department would first type data into a spreadsheet, then manually plot that data onto a map, point by point.

Not without limitations

General Utility Package (GUP), the UNIX-based system Concert used, was designed for combining electronic route position lists and charts for submarine cables. The system was proprietary and could not easily import off-the-shelf coastline data, bathymetry data, or other cable routes unless this information was directly digitized into the system—a time-consuming process. Likewise, the GUP was not able to export data into a format that others, like survey companies, installation companies, or the Coast Guard, could use—so all data exchange was carried out by fax machine. This worked relatively well until other submarine cable data users began using GIS. Once they began operating in a digital environment, these companies quickly developed a preference for electronic data entry over manual entry.

Concert then approached Thales Geo-Solutions (Pacific), Inc., formerly known as Racal Pelagos, Inc., of San Diego, California, to assist with modernizing its cable data management systems.

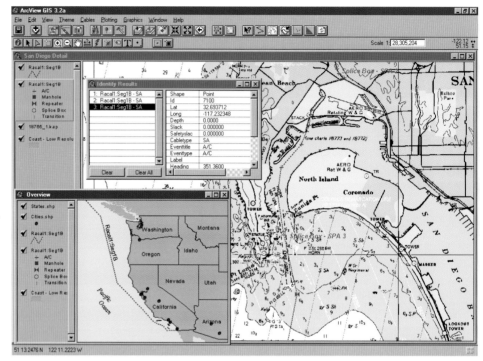

Users better understand the system as a whole when they can see an overview of the cable network and a detailed view of its environment. The Identify Results table gives details about specific cable segments.

Moving to GIS

In 1987, software developed by Thales GeoSolutions (Pacific) was used onboard the C. S. Long Lines to plan the installation of the world's first transatlantic fiber-optic cable network. Since that time, Thales GeoSolutions (Pacific) has been providing software solutions to companies like Concert for both office and maritime applications.

By 1998, the two companies began developing a system to move Concert's data records management to a PC-based GIS environment. The result was Cable Analyst™, an extension that works with ArcView GIS to provide several import, editing, printing, and charting tools for the submarine cable industry.

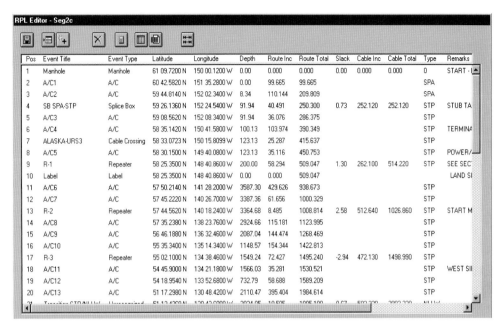

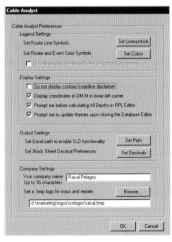

ArcView GIS extension Cable Analyst provides an interface for both entering and editing data. The data formats output from Cable Analyst are somewhat standardized within the submarine cable industry and its related fields.

One step at a time

Thales GeoSolutions (Pacific) began building Cable Analyst as an ArcView GIS-based charting system. The primary goal was to provide a basic charting tool that would automatically generate "north-up" maps along cable routes. At the same time, the company worked to convert Concert's data to a standardized format that was compatible with the GIS system.

Once that was done, the company was able to add tools for loading, editing, printing, and querying submarine cable routes. Also added were tools for interactive route development and planning, such as a Great Circle guideline tool that provides the shortest course between two points on the earth. This is especially important for global-scale fiber-optic cable routes like those managed by Concert.

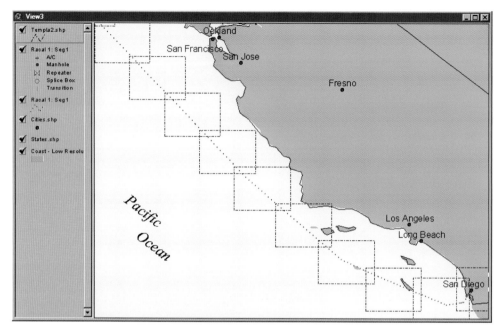

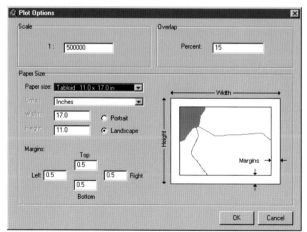

The boxes on the map above show a series of north-up maps automatically created for an entire segment with just a few button clicks in a custom dialog box like the one at the left.

Providing the tools for success

Cable Analyst provides an Import Wizard through which users can view data as it is imported. Once a route has been entered into the system, it is stored in a Microsoft Access database. From there it can easily be converted to an ArcView shapefile whenever a user wants to view or edit it.

Cable Analyst has a variety of custom tools for editing these database files. Users can copy and delete records, or alter their position. As changes are made to the network, new records are added to the database. Cable Analyst ensures that any new addition to the network, such as a splice box, is added both in the correct sequence along the cable and at the correct latitude and longitude.

Using code written with ESRI's Avenue scripting language, Cable Analyst lets cable operators accurately display locations in decimal degrees without having to perform tedious computations themselves.

Although it has yet to import all of its existing routes into Cable Analyst, Concert now uses this software to plan new routes and to perform other charting functions.

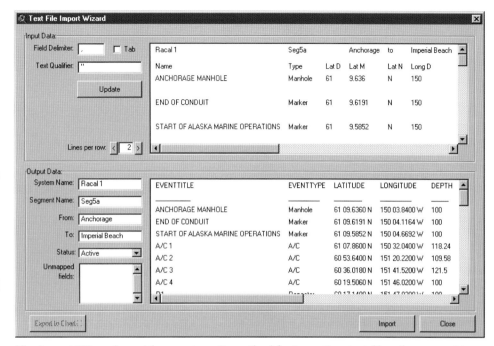

The Import Wizard provides a systematic method for importing text files of route position lists (RPLs) from various sources into the Cable Analyst database format. Concert can add attribute data, such as owner information, installation date, and maintenance records, as well as current cable status (active, retired, or planned) for the routes in the system.

Adding to the data

Laying new submarine cable involves more than just knowing where existing cables are; it means understanding the temperature, ocean currents, and depth of the ocean at various points, and how these conditions can shift the cable as it falls, so that it ends up somewhere near where it's supposed to.

Cable Analyst comes with coastline data, both low- and high-resolution, as well as underwater bathymetry contour data from the British Oceanographic Data Center. Analysts can use the coastline data and bathymetry contours to display, plot, or calculate depths in a wide range of map scales. Other relevant data can be easily loaded into

ArcView, including electronic nautical charts. Being able to view this data as a map proves invaluable at times when, for example, network analysts need to see more than one cable system at a time in order to record cable crossings.

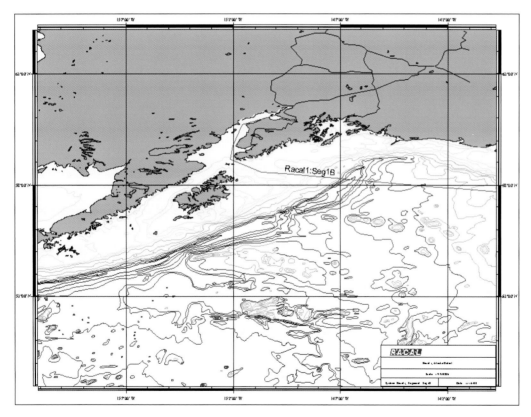

Simple menus make it easy to generate maps like this showing the path of an underwater cable.

Sharing information

Because the U.S. Coast Guard may need one format while a local survey company may need another, it was important for Concert to find a way to create output in a variety of formats.

In addition to maps shared with seafaring vessels, Cable Analyst lets Concert produce route position lists in industry-standard Block Sheets and Straight Line Diagrams that are used by planners, engineers, and all the people who make, install, or repair cables.

Cable Analyst also provides special tools that let engineers chart routes and create layouts at a specified scale along the entire route, something they were unable to do with their original software.

The long-standing relationship between Concert and Thales will continue to provide developments and enhancements to the Cable Analyst product, ensuring that both concerns will remain at the industry forefront.

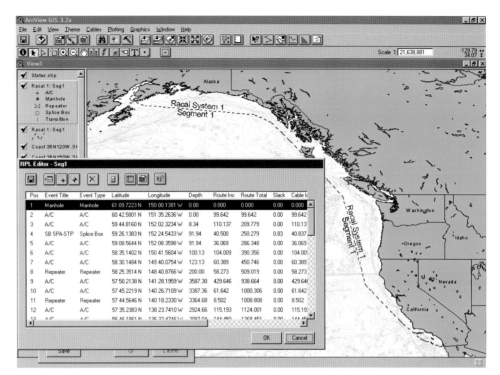

Users can edit, delete, add, print, and calculate data using the route position lists editor, which displays the tabular version of a route.

The system

733-MHz HP® PC with 256 MB RAM and 20-GB hard drive, 21-inch Nokia® monitor, and HP 1055CM color plotter.

Software: ArcView, Microsoft Office 97, Windows NT, CalComp® digitizer, CalComp scanner

The data

The fictional data used for the screen captures is included in the Cable Analyst Tutorial.

The people

Thanks to David Millar and Alec Bost of Thales GeoSolutions (Pacific), Inc., and James Murray and Gino Montes of Concert.

Thales GeoSolutions (Pacific), Inc., was formerly known as Racal Pelagos, Inc.

GIScience

GIS for Everyone SECOND EDITION

Now everyone can create smart maps for school, work, home, or community action using a personal computer. This revised second edition includes the ArcExplorer™ geographic data viewer and more than 500 megabytes of geographic data. ISBN 1-879102-91-9 196 pages

The ESRI Guide to GIS Analysis, Volume 1: Geographic Patterns and Relationships

An important new book about how to do real analysis with a geographic information system. *The ESRI Guide to GIS Analysis* focuses on six of the most common geographic analysis tasks. ISBN 1-879102-06-4 188 pages

Modeling Our World: The ESRI Guide to Geodatabase Design

With this comprehensive guide and reference to GIS data modeling and to the new geodatabase model introduced with ArcInfo™ 8, you'll learn how to make the right decisions about modeling data, from database design and data capture to spatial analysis and visual presentation. ISBN 1-879102-62-5 216 pages

Hydrologic and Hydraulic Modeling Support with Geographic Information Systems

This book presents the invited papers in water resources at the 1999 ESRI International User Conference. Covering practical issues related to hydrologic and hydraulic water quantity modeling support using GIS, the concepts and techniques apply to any hydrologic and hydraulic model requiring spatial data or spatial visualization. ISBN 1-879102-80-3 232 pages

Beyond Maps: GIS and Decision Making in Local Government

Beyond Maps shows how local governments are making geographic information systems true management tools. Packed with real-life examples, it explores innovative ways to use GIS to improve local government operations. ISBN 1-879102-79-X 240 pages

The ESRI Press Dictionary of GIS Terminology

The *ESRI Press Dictionary of GIS Terminology* brings together the language and nomenclature of the many GIS-related disciplines and applications. Designed for students, professionals, researchers, and technicians, the dictionary provides succinct and accurate definitions of more than a thousand terms. ISBN 1-879102-78-1 128 pages

Planning Support Systems: Integrating Geographic Information Systems, Models, and Visualization Tools

Richard Brail of Rutgers University's Edward J. Bloustein School of Planning and Public Policy, and Richard Klosterman of the University of Akron, have assembled papers from colleagues around the globe who are working to expand the applicability and understanding of the top issues in computer-aided planning. ISBN 1-58948-011-2 468 pages

Geographic Information Systems and Science

This comprehensive guide to GIS, geographic information science (GIScience), and GIS management illuminates some shared concerns of business, government, and science. It looks at how issues of management, ethics, risk, and technology intersect, and at how GIS provides a gateway to problem solving, and links to special learning modules at ESRI Virtual Campus (campus.esri.com). ISBN 0-471-89275-0 472 pages

Mapping Census 2000: The Geography of U.S. Diversity

Cartographers Cynthia A. Brewer and Trudy A. Suchan have taken Census 2000 data and assembled an atlas of maps that illustrates the new American diversity in rich and vivid detail. The result is an atlas of America and of Americans that is notable both for its comprehensiveness and for its precision. ISBN 1-58948-014-7 120 pages

ESRI Map Book, Volume 16: Geography—Creating Communities

A full-color collection of some of the finest maps produced using GIS software. Published annually since 1984, this unique book celebrates the mapping achievements of GIS professionals. *Directions Magazine* (www.directionsmag.com) has called the *ESRI Map Book* "The best map book in print." ISBN 1-58948-015-5 120 pages

CONTINUED ON NEXT PAGE

Other books from ESRI Press continued

The Case Studies Series

ArcView GIS Means Business

Written for business professionals, this book is a behind-the-scenes look at how some of America's most successful companies have used desktop GIS technology. The book is loaded with full-color illustrations and comes with a trial copy of ArcView GIS software and a GIS tutorial. ISBN 1-879102-51-X 136 pages

Zeroing In:

Geographic Information Systems at Work in the Community

In twelve "tales from the digital map age," this book shows how people use GIS in their daily jobs. An accessible and engaging introduction to GIS for anyone who deals with geographic information. ISBN 1-879102-50-1 128 pages

Serving Maps on the Internet

Take an insider's look at how today's forward-thinking organizations distribute map-based information via the Internet. Case studies cover a range of applications for ArcView Internet Map Server technology from ESRI. This book should interest anyone who wants to publish geospatial data on the World Wide Web. ISBN 1-879102-52-8 144 pages

Managing Natural Resources with GIS

Find out how GIS technology helps people design solutions to such pressing challenges as wildfires, urban blight, air and water degradation, species endangerment, disaster mitigation, coastline erosion, and public education. The experiences of public and private organizations provide real-world examples. ISBN 1-879102-53-6 132 pages

Enterprise GIS for Energy Companies

A volume of case studies showing how electric and gas utilities use geographic information systems to manage their facilities more cost effectively, find new market opportunities, and better serve their customers. ISBN 1-879102-48-X 120 pages

Transportation GIS

From monitoring rail systems and airplane noise levels, to making bus routes more efficient and improving roads, this book describes how geographic information systems have emerged as the tool of choice for transportation planners. ISBN 1-879102-47-1 132 pages

GIS for Landscape Architects

From Karen Hanna, noted landscape architect and GIS pioneer, comes *GIS for Landscape Architects*. Through actual examples, you'll learn how landscape architects, land planners, and designers now rely on GIS to create visual frameworks within which spatial data and information are gathered, interpreted, manipulated, and shared. ISBN 1-879102-64-1 120 pages

GIS for Health Organizations

Health management is a rapidly developing field, where even slight shifts in policy affect the health care we receive. In this book, you'll see how physicians, public health officials, insurance providers, hospitals, epidemiologists, researchers, and HMO executives use GIS to focus resources to meet the needs of those in their care. ISBN 1-879102-65-X 112 pages

GIS in Public Policy

This book shows how policy makers and others on the front lines of public service are putting GIS to work—to carry out the will of voters and legislators, and to inform and influence their decisions. *GIS in Public Policy* shows vividly the very real benefits of this new digital tool for anyone with an interest in, or influence over, the ways our institutions shape our lives. ISBN 1-879102-66-8 120 pages

Integrating GIS and the Global Positioning System

The Global Positioning System is an explosively growing technology. *Integrating GIS and the Global Positioning System* covers the basics of GPS and presents several case studies that illustrate some of the ways the power of GPS is being harnessed to GIS, ensuring, among other benefits, increased accuracy in measurement and completeness of coverage. ISBN 1-879102-81-1 112 pages

GIS in Schools

GIS is transforming classrooms—and learning—in elementary, middle, and high schools across North America. *GIS in Schools* documents what happens when students are exposed to GIS. The book gives teachers practical ideas about how to implement GIS in the classroom, and some theory behind the success stories. ISBN 1-879102-85-4 128 pages

Disaster Response: GIS for Public Safety

GIS is making emergency management faster and more accurate in responding to natural disasters, providing a comprehensive and effective system of preparedness, mitigation, response, and recovery. Case studies include GIS use in siting fire stations, routing emergency response vehicles, controlling wildfires, assisting earthquake victims, improving public disaster preparedness, and much more. ISBN 1-879102-88-9 136 pages

Open Access: GIS in e-Government

A revolution taking place on the Web is transforming the traditional relationship between government and citizens. At the forefront of this e-government revolution are agencies using GIS to serve interactive maps over their Web sites and, in the process, empower citizens. This book presents case studies of a cross-section of these forward-thinking agencies. ISBN 1-879102-87-0 124 pages

GIS in Telecommunications

Global competition is forcing telecommunications companies to stretch their boundaries as never before—requiring efficiency and innovation in every aspect of the enterprise if they are to survive, prosper, and come out on top. The ten case studies in this book detail how telecommunications competitors worldwide are turning to GIS to give them the edge they need. ISBN 1-879102-86-2 120 pages

Conservation Geography: Case Studies in GIS, Computer Mapping, and Activism

This collection of dozens of case studies tells of the ways GIS is revolutionizing the work of nonprofit organizations and conservation groups worldwide as they rush to save the earth's plants, animals, and cultural and natural resources. As these pages show clearly, the power of computers and GIS is transforming the way environmental problems and conservation issues are identified, measured, and ultimately, resolved. ISBN 1-58948-024-4 252 pages

CONTINUED ON NEXT PAGE

Other books from **ESRI Press** continued

ESRI Software Workbooks

Understanding GIS: The ARC/INFO® Method (UNIX®/Windows NT® version)

A hands-on introduction to geographic information system technology. Designed primarily for beginners, this classic text guides readers through a complete GIS project in ten easy-to-follow lessons. ISBN 1-879102-01-3 608 pages

Understanding GIS: The ARC/INFO Method (PC version)

ISBN 1-879102-00-5 532 pages

ARC Macro Language:
Developing ARC/INFO Menus and Macros with AML

ARC Macro Language (AML™) software gives you the power to tailor ARC/INFO Workstation software's geoprocessing operations to specific applications. This workbook teaches AML in the context of accomplishing practical ARC/INFO Workstation tasks, and presents both basic and advanced techniques. ISBN 1-879102-18-8 826 pages

Getting to Know ArcView GIS

A colorful, nontechnical introduction to GIS technology and ArcView GIS software, this workbook comes with a working ArcView GIS demonstration copy. Follow the book's scenario-based exercises or work through them using the CD and learn how to do your own ArcView GIS project. ISBN 1-879102-46-3 660 pages

Extending ArcView GIS

This sequel to the award-winning *Getting to Know ArcView GIS* is written for those who understand basic GIS concepts and are ready to extend the analytical power of the core ArcView GIS software. The book consists of short conceptual overviews followed by detailed exercises framed in the context of real problems. ISBN 1-879102-05-6 540 pages

Getting to Know ArcGIS Desktop: Basics of ArcView, ArcEditor, and ArcInfo

Getting to Know ArcGIS Desktop is a workbook for learning ArcGIS™, the newest GIS technology from ESRI. Readers learn to use the software that forms the building blocks of ArcGIS: ArcMap™, for displaying and querying maps; ArcCatalog™, for managing geographic data; and ArcToolbox™, for setting map projections and converting data. Richly detailed illustrations and step-by-step exercises teach basic GIS tasks. Includes a fully functioning 180-day trial version of ArcView 8 software on CD–ROM, as well as a CD of data for working through the exercises. ISBN 1-879102-89-7 552 pages

ESRI educational products cover topics related to geographic information science, GIS applications, and ESRI technology. You can choose among instructor-led courses, Web-based courses, and self-study workbooks to find education solutions that fit your learning style and pocketbook. Visit www.esri.com/education for more information.

ESRI Press publishes a growing list of GIS-related books. Ask for these books at your local bookstore or order by calling 1-800-447-9778. You can also shop online at www.esri.com/gisstore. Outside the United States, contact your local ESRI distributor.

ESRI Press • 380 New York Street • Redlands, California 92373-8100 • www.esri.com/esripress